W9-CCE-518

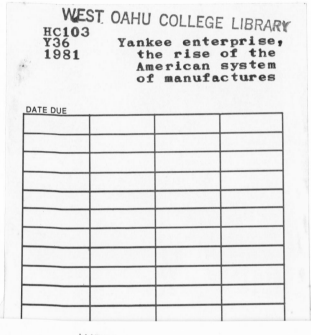

Yankee Enterprise

The Rise of the American System of Manufactures

Otto Mayr and Robert C. Post

Editors

A Symposium Sponsored by the
United States Chamber of Commerce
held at the
Dibner Rare Book Library
National Museum of American History
Smithsonian Institution

Smithsonian Institution Press
Washington, D.C., 1981

Cover: The Wheeler and Wilson ma-
chineshop, Bridgeport, Connecticut.
(From Charles H. Fitch, "Report on
the Manufactures of Interchangeable
Mechanism," in *Report on the Manufactures of the United States at the
Tenth Census* [Washington, D.C.,
1883]; Smithsonian Institution Neg.
No. 81-206.)

Library of Congress Cataloging in Publication Data

Yankee enterprise, the rise of the American system of manufactures.

Bibliography: p.
Includes index.
1. United States—Industries—History—Congresses. 2. Industrial arts—
United States—History—Congresses. 3. Industrial management—United
States—History—Congresses. 4. Interchangeable mechanisms—Congresses.
I. Mayr, Otto. II. Post, Robert C. III. Chamber of Commerce of the United
States. IV. Dibner Library. HC103.Y36   338.0973   81-607315
ISBN 0-87474-634-5 (cloth)   AACR2
ISBN 0-87474-631-0 (pbk.)

CONTENTS

FOREWORD

In the essays that follow, distinguished historians and scholars present thought-provoking, sometimes divergent ideas on "Yankee enterprise"—how the rise of the American system of manufactures was shaped by, and in turn contributed to, American society. This society, as Otto Mayr and Robert Post note in their introduction, was one in which "individual initiative was encouraged by a host of factors, and official restrictions were anathema to ordinary citizens."

The U.S. Chamber of Commerce is pleased to sponsor publication of this unique volume, which compiles a series of papers first presented at the Smithsonian Institution's symposium entitled "The Rise of the American System of Manufactures."

We are proud to play a role in adding to knowledge of the history and effects of American mass production. This significant topic has been the object of far too little analysis and discourse in the past; *Yankee Enterprise* represents a major step toward correcting previous neglect. It will prove to be a valuable and comprehensive addition to the study of American industry and culture.

Richard L. Lesher
President,
Chamber of Commerce of the United States

ACKNOWLEDGMENTS

Americans have always been notably fond of establishing organizations aimed at advancing enterprises of mutual interest. Organizations of businessmen, usually called Boards of Trade, date back to the founding of the republic. Their proliferation eventually resulted in 1912 in the formation of a national organization called the Chamber of Commerce of the United States of America. Today it has more than 125,000 members, some of them state and local chambers of commerce and trade and professional associations, but the majority, over 95 percent, individual business firms. The U.S. Chamber of Commerce has a keen interest in the technological, managerial, and promotional innovations that undergird modern business enterprise. That interest is certified by its generous sponsorship of the symposium from which this book derives, and for which I want to extend a personal word of thanks.

Thanks are also due to one particular member of the U.S. Chamber of Commerce, the Burndy Corporation of Norwalk, Connecticut. This firm was founded in 1924 by Bern Dibner, on the basis of several patents for devices to connect electrical cables. Under Dibner's leadership it grew into a prime manufacturer of electrical and electronics apparatus, with plants in nine nations besides the United States. Dibner himself, having witnessed the growth of an industry from the germ of a simple idea, became interested in the origins of electrical technology and ultimately in the general history of science and technology. Byproducts of his own researches are two major research institutions, the Burndy Library in Norwalk and the Dibner Library at the National Museum of American History in Washington, D.C.

The heart of the Dibner Library is a collection of more than 8,000 printed books, classics in the history of science and technology, which draw scholars from all corners of the globe. The library's formal programs include symposia on significant aspects of the history of science and technology. It strikes me as altogether fitting and proper that the first of these symposia should treat the American system of manufactures. For here is a theme central not only to the history of American business enterprise and the history of American technology, but to the whole broad panorama of American culture, the reflection and interpretation of which is the central mission of the National Museum of American History.

Roger G. Kennedy
Director,
National Museum of American History

INTRODUCTION

Since the Renaissance one revolution after another—religious, political, social, industrial—has upset the prevailing order. Yet no revolution has wrought changes more profound than the advent of mass production. In some parts of the world the mass-production revolution has effaced class distinctions and homogenized international differences; in other parts, by encouraging rising expectations when fulfillment remains problematic, its impact has been terribly divisive. Even where mass production has bestowed comforts and conveniences that no king could have granted two centuries ago, nobody regards it with any genuine affection. And there is a pervasive irony: its most outspoken critics are Westerners, the beneficiaries of its richest bounties.

The mass-production revolution was fomented chiefly in "the first new nation," the United States of America, though there is no question that its seeds were planted in the Old World. Some of the infrastructure, notably the factory system and division of labor, originated in Europe. And some of the technological principles had even been put into practice there. Still, in the context of the Old World these circumstances seem anomalous, while in America, by contrast, techniques of quantity production—and extension of these techniques to an ever broader range of products—became fundamentals in the nation's social and economic history. Indeed, in an important sense these techniques also became fundamental to the nation's political history and to the unmatched resiliency of its governmental institutions. As David Potter observed in his study of the American character, *People of Plenty:* "European radical thought is prone to demand that the man of property be stripped of his carriage and his fine clothes. But American radical thought is likely to insist, instead, that the ordinary man is entitled to mass-produced copies, indistinguishable from the originals."

Pioneers in this American saga of "mass-produced copies" include Samuel Slater, Oliver Evans, Jacob Perkins, Thomas Blanchard, Eli Terry, and the controversial Eli Whitney, who promised more than he delivered in the way of making musket parts interchangeable, but whose role as a propagandist for innovative production concepts must not be underrated. By the middle of the nineteenth century, American entrepreneurs had attained high-quantity outputs not only of small arms but also of other precision articles from locks to machine tools, not to mention machinery for which precision fits were not crucial, such as reapers and steam engines.

Some of these products were marketed overseas even in the 1840s, and by the next decade the author of a primer for German immigrants could refer with no undue exaggeration to the capability of the United States "to compete with the manufacture of even the greatest and oldest industrial countries in the world." Europe was served truly dramatic notice of technological attainments in the one-time colonial lands in 1851, at the "Great Exhibition of the Industry of All Nations," staged in London in a "Crystal Palace" that became the archetype for several others elsewhere, including New York two years later. The production machinery and products exhibited in London by American entrepreneurs such as Alfred Hobbs, Samuel Colt, and Cyrus McCormick created a sensation. Hobbs could turn out locks, Colt could turn out revolvers, and McCormick could turn out reapers in quantities that seemed incredible. The efficiency of McCormick's machines left Britons dumbfounded. Rifles exhibited by the Vermont firm of Robbins & Lawrence actually could be disassembled, then satisfactorily put back together from jumbled components—thus dispelling tinges of the apocryphal from tales of that feat once and for all.

A commission of British experts was promptly dispatched to investigate American production technology at first hand. The commissioners observed numerous processes that involved "adaptation of special apparatus to a single operation," and soon the phrase "American system of manufactures" had attained currency. English entrepreneurs tended to be slow to see its virtues, but in America the system was applied ever more widely and with the utmost enthusiasm. The heart of the system when the British first took notice was the sequential arrangement of single-purpose machine tools. Though this greatly accelerated the rate of parts production, rarely did these tools turn out parts that were fully interchangeable, not when the finished product required any appreciable degree of precision fitting. Interchangeable parts remained an exotic variant on the new American production technology in the antebellum period.

After the Civil War, manufacturers of precision-fit articles began to strive for a degree of uniformity that would permit assembly from components selected at random, and the 1880 census report included an essay on the "general growth of the 'interchangeable' system of manufacturing."

Samuel Colt's display at the Crystal Palace, New York, 1853. (Smithsonian Institution Neg. No. 81–204.)

The author, Charles Fitch, thought that "it may not be too much to say that, in some respects, this system has been one of the chief influences in the rapid growth in the national wealth." This observation is likely to strike a present-day reader—familiar with Henry Ford's prodigious impact on American economic history—as too timid. Yet even in 1880, interchangeability remained rather unusual. What is remarkable is that certain industries had attained a capability for enormous outputs of articles that still were hand fitted in the end. It seemed clear as long as a century ago that the United States was on its way to becoming the premier exporter of manufactured goods and the wealthiest nation on earth. But the point to reiterate in the present context is that interchangeable parts were *not* an essential concomitant of high-quantity production. Nor was the latter necessarily synonymous with mass production, though here the distinction becomes sufficiently subtle to demand further explanation.

What we now think of as mass production was well established in all its fundamentals by the time the Guns of August signalled the significance it would have for the style of modern warfare. Other nations had mastered the techniques, too—for capital goods if not necessarily for consumer durables—and history had come an ironic full circle. For it was in the realm of arms manufacture that the first attempts to attain interchangeability of parts had been made at the turn of the nineteenth century. When it became clear early in the twentieth that techniques existed for keeping armies supplied with limitless munitions, people first called to question the assumption that the mass-production revolution was inherently beneficent.

Critics of mass production's social costs speak with increasing urgency as the twentieth century draws to a close. Yet if one rejects the notion that technology is in any sense autonomous, then the calamities of our age can be attributed only to mankind's intrinsic imperfections or to persisting imperfections in systems of social control. Indeed, as revolutions go, the revolution wrought by mass production *is* benign. Unlike political revolutions, in which irrational forces are always a component and often predominant, every step has necessitated rational planning. While there are key epochs that can be dated with some precision, this revolution had no 1776, no 1789, and actually it is a cumulative process that continues into our own day.

It seems certain that no other concatenation of events has so profoundly altered social structures and economic functions—not since the Neolithic revolution when Stone Age nomads turned to a settled existence based on agriculture and fixed communities. Yet general historians have scarcely begun to consider mass production's constituent elements, and most of them are still at the level of repeating what Eugene Ferguson calls "jejune folklore." Among historical specialists, while none has yet produced a comprehensive interpretation of that element we call the American system of manufacturers—or even a telling heuristic thrust—there is a growing collegium of scholars devoted to the subject, and their accomplishments are not inconsiderable. They have taken us beyond the thrall of simplistic theoretical models, and they have introduced us to the evidence that factors such as management and marketing have been as crucial to American preeminence in mass production as has technology per se. "As each author has followed his particular star," Ferguson writes, "the territory encompassed by the American system of manufactures has expanded, and we are no longer talking only about the armory of Eli Whitney, but about the many elements of a larger, indeed world-wide, phenomenon of machine-based mass production for mass markets."

In March 1978 a symposium was held at the National Museum of American History, with the purpose of convening the most active members of this collegium to tell what they had been learning as each followed his particular star. This seemed a promising venture since the collegium is otherwise heterogeneous, and if we know anything at all about this complex phenomenon we know that a mature understanding demands a synthesis of viewpoints. Among the papers presented were four by scholars in the history of technology, two by economic historians, and one each by a labor historian, a business historian, and a social historian. Reviewed in accord with critiques developed and elaborated during the symposium, these papers are what follow.

This publication has three primary objectives. By epitomizing current scholarship, it may serve as a useful reference. By underscoring the historical significance of the American system of manufactures, it may engender a more thoughtful treatment in general historical narratives. And by juxtaposing the interpretive and conceptual inconsistencies bound to emerge from any symposium, it may challenge specialists to ponder the dimensions of the comprehensive synthesis so urgently needed.

The essays tend to concentrate on origins and early development, though in many instances they carry past the end of the nineteenth century, and the final one is focused on our own century. Anyone attempting a synthesis will be sternly tested by the effort to sort out the story up to the point when Europeans first noted the novel character of American production technology. The story could be pursued, nevertheless, into the 1920s or even further, to encompass such exotic permutations as numerical control and full-blown automation.

Three phrases are central to all of the following discussion: "American system of manufactures," "mass production," and "production with interchangeable parts." It may be emblematic of the state of scholarship within this specialty that disagreement persists about the definition of these phrases, especially the first one. Some participants in the symposium equate it with the entire saga of American mass production. Others adhere to (or imply adherence to) a more limited definition. And while it is obviously incumbent upon the editors to pay some attention to the matter of definition, the fact that we have not yet done so—indeed have bandied the phrase about rather carelessly—suggests that here are disagreements not readily resolved.

A case may be made for the American system of manufactures being simply the novel shop techniques the British observers saw in the 1850s. It goes without saying that this was not the whole of what was later dubbed mass production. Yet to narrow the definition of the American system of manufactures too much tends to exclude the most revealing insights of recent scholarship. Daniel Nelson points out, for example, that entrepreneurs who introduced or adopted the system "installed more than a series of machines and a system of machine production." Any such entrepreneur was also compelled to ponder "fundamental decisions regarding treatment of his labor force, his relations with his subordinates, and, ultimately, his conception of social progress."

Nelson sees the system as congruent with manufactures that required a substantial proportion of skilled workers and entailed little of the "dirty" work that prevailed in capital-goods industries. Paul Uselding, thinking along the same lines, focuses his attention on refinements in the processes of machining metallic components. Alfred Chandler is in accord with this approach to the subject insofar as "techniques for designing and managing the movements of men" were given closest attention in those industries based on precision machinework. Chandler hardly presumes, however, that the American system of manufactures was confined to these industries, or even to industries that produced consumer goods. Both he and Albert Musson consider the subject as germane to heavy industry as well as light, and both seem to consider it as synonymous with mass production.

Though Chandler and Musson each present a convincing argument, the question remains open. Moreover, there is a fundamental conceptual disparity between these two senior scholars. In marshalling evidence for the British origins of what was later dubbed the "American" system, Musson implies that tools as such were its key component. Chandler, on the other hand, emphasizes the complex make-up of the system, especially in its maturing stages: it was the interactions between production technology, the organization of the labor force, and the administrative demands of cost accounting and of a farflung distribution system that together gave rise to the management techniques upon which all large industries are now based.

Despite leaving key questions unresolved, participants in the symposium did make some progress toward a shared definition. Most agreed, at least implicitly, that the system was something transcending mere technology, something suggestive of an underlying managerial philosophy upon which the particulars were based. Most agreed that ultimately the system embodied an accretion of elements. But they did not always agree about which elements.

Mass production in its modern guise includes the factory as a logistic concept, division of labor as an organizational concept, the special-purpose machine tool as a technological concept, scientific management as an intellectual concept, and fabrication from interchangeable parts as a concept at once logistical, organizational, technological, and abstract. Promotional strategies and methods of mass distribution also are central. But which of these aspects of mass production can likewise be considered as part of the American system of manufactures? There is nothing wrong with endowing the phrase with an evolutionary dimension, as Ferguson does, and treating the two as synonymous. Among those who choose to regard the system as a nineteenth-century precursor of mass production, a twentieth-century phenomenon, there are, however, basic disagreements. When considering the essays that follow, it soon becomes clear that the most ubiquitous of them centers on the matter of interchangeability. This disagreement is rarely explicit, in the way (say) that Musson and Nathan Rosenberg clash head-on over the origins of the system, each according to his own definition. It is just that certain of our participants presume an essential congruence of the system with interchangeable parts, while others deny it.

Much of this has to do with divergent preconceptions that the authors do not always specify. Yet it is intriguing that, even with the spotlight on so many altogether fresh dimensions of the system, the most obvious disagreement should turn on such a traditional point of departure. We have no intention of trying to resolve this dilemma here, though it does bear reiterating that interchangeability was not in itself essential to the system that the British observers reported on in the 1850s. Quantity production of a complex mechanism with such precision that components could be stockpiled at random and assembled without hand fitting—here was a technique available for adoption, but, at least in the antebellum years, it was more likely to increase costs than decrease them.

This point is strongly emphasized by Ferguson, but it is Ferguson's student David Hounshell who specifies the distinction that logically follows—the distinction between the American system of manufactures in general and "New England armory practice" in particular. Interchangeability was obviously a useful feature when field repairs were required, and, of the manufacturers visited by the British observers of the 1850s whose products had a "thorough identity of parts," all were armsmakers. Otherwise, the primary incentive to adopt the American system of manufactures, reduced production costs, was best served without attempting to attain thoroughgoing interchangeability. This, of course, did not apply to manufactures that required only a low degree of precision, such as wooden clocks. It applied most strongly to manufactures with very close tolerances, and some concerns—Singer above all—attained an enormous output while still employing legions of filers and hand fitters.

It seems reasonably clear, then, that no hard-and-fast equation can be drawn between the American system and interchangeable parts. And, despite the capabilities of a firm such as Singer, there is even a good question regarding its equivalence to mass production. Read, for example, the words of the man who did more than anyone to put the term "mass production" into everyday usage. In an article prepared for the thirteenth edition of the *Encyclopaedia Britannica*, published in 1926, Henry Ford wrote:

> Mass production is not merely quantity production, for this may be had with none of the requisites of mass production. Nor is it merely machine production, which also may exist without any resemblance to mass production. Mass production is the focussing upon a manufacturing system of the principle of power, accuracy, system, continuity, and speed.

"Power, accuracy, system, continuity, and speed"—here are five separate and distinct facets of the production process, though Ford's insistence on distinguishing between quantity production and mass production still strikes some historians as akin to a debate over angels on pinheads. Certainly most textbook definitions of mass production still equate it with production in quantity by machinery, nothing more, just as they equate the American system with fabrication from interchangeable parts. General historians seem to have no qualms about declaring, as Ferguson laments, that Ford's accomplishment was "to adapt to car manufacturing the production techniques first developed by small arms manufacturers."

If this is fundamentally erroneous, then where does that leave the American system of manufactures? Is it properly something peculiar to the middle nineteenth century, one stage in the evolution of manufacturing techniques, a stage that just happened to be reported on by some technically astute visitors from Britain? Perhaps; though even Hounshell, one of the stoutest proponents of a limited definition, lends support to the view that it must not be defined *too* narrowly. Each of three different sewing-machine manufacturers, he shows, developed a distinct variant on "the system."

A good many vexed issues require as rigorous an analysis as this one does. Like any important historical phenomenon, the rise of the American system of manufactures may be viewed from diverse perspectives. In the chapters that follow, certain facets of this phenomenon—technological transfer, resource endowment, and governmental patronage, for instance —are considered in depth. But many others are scarcely touched on. And besides the questions left unanswered, it seems certain that more are neither answered nor asked. The American system of manufactures was the creation of a remarkable generation of inventor-entrepreneurs. We know little enough about most of them as individuals; more importantly, we know next to nothing about the ways they interacted. Did they share similar formative experiences? Did they dream the same dreams?

On the question of some shared spiritual environment, it can hardly be irrelevant that the American system of manufactures took root in a society infused with the tenets of classical liberalism, a society in which individual initiative was encouraged by a host of factors and official restrictions were anathema to ordinary citizens. When the government got involved, it was either as a partner or as a benefactor. This did not happen much, but when it did happen, in the armsmaking industry, the consequences were very significant indeed. As Merritt Roe Smith puts it:

> The introduction of new tools and machinery and the consequent specialization of labor in the arms industry engendered a situation that required greater coordination and control of the work process and . . . military traditions within the Ordnance Department encouraged and fostered the use of paternalistic controls as a means of imposing order in the workplace.

Here is evidence aplenty that the American system of manufactures had a far broader range of manifestations than mere technology. As for technology *qua* technology, the infusion of public money was perhaps a necessary though not sufficient condition for the inception of the technology of interchangeable parts. Otherwise the impetus behind the new production technology lay in conventional entrepreneurial urges. Perhaps the conceptual cornerstone was that subtle notion labeled "economies of scale."

The label is rather inadequate, but the concept is clear enough. It entails a closed-loop argument that begins with a comprehension that high-volume output requires specialized machinery, much of it designed to perform only one operation, hence a large initial investment. The second segment of the loop is predicated on the assumption that the market for a product will grow as the unit cost falls, and the third on the assumption that overall profits will increase even as per-unit profits decrease. Closing the loop entails the assumption that these profits will offset the large initial investment in production facilities.

The initial investment requires a leap of faith—faith in the three successive assumptions. Such a leap is not for the timid. But those who leaped—were they conscious of the rationale we have set forth and were they spurred on by it? If so, where did this concept originate, and how was it propagated? To consider such a question is to consider something that has nothing to do with technology or economics, but rather belongs to the history of ideas.

Likewise, when Neil Harris turns his attention to "object consciousness" and the "American system of consumption," showing how the technological characteristics of the system affected the thinking of those who partook of its products, he too serves notice that we will never grasp the system's full significance so long as we confine ourselves to economic history or the history of technology—or, indeed, so long as we confine research to traditional historical sources. Harris's sources are primarily works of fiction that relate the buying process to the social experience.

That is what this book is primarily about: "social experience." The traditional "tool centered" way of thinking about the American system of manufactures, while not absent here, is scarcely predominant either. Yet despite the presence of several new approaches to the subject, we concede that a number of questions remain unexplored and a number of ambiguities remain unresolved. On the way to a mature understanding of the American system of manufactures, this volume can be only a step. But even one step is important. For if we are correct in asserting that the American system of manufactures lies at the epicenter of one of the most profound revolutions in history, then the need to achieve a mature understanding must be one of our primary intellectual priorities.

Otto Mayr
Robert C. Post

HISTORY AND HISTORIOGRAPHY

Eugene S. Ferguson

In a campaign launched in 1851 and extending over three generations, American technology conquered hearts and markets throughout the world. Singer sewing machines, Kodak cameras, and Ford automobiles came to symbolize the American cornucopia that poured out consumer durables in unprecedented quantities. Governments and individuals might hate America and the Americans, but they coveted American technology and willingly mortgaged their futures to enjoy its charms. Since World War II, American monopolies have been challenged by Japanese and European entrepreneurs, but the new competitors have merely adopted the successful American formulas of mass production and mass marketing for mass consumption.

Whatever one may think of the social and environmental consequences of the American triumph, it was on its own terms immensely successful. It provided the blueprint for modern industrial development, and no other system has proved to be as bountiful and dependable as a producer of material wealth. Since early in the nineteenth century, American products have been made according to the "American system of manufactures," a term coined by an English observer in the 1850s and adopted by historians in the 1930s.[1] Under the American system, the English notion of minute division of labor was extended and transformed first by mechanization and later by systematic control of work and workers. The immediate result was an abundance of products in quantities limited only by the market's ability to absorb them.

Despite the powerful role that the American system of manufactures has played as a worldwide agent of social change, its history has been esoteric in the extreme. Historians of technology have been fascinated by the quest within American armories for small arms having interchangeable parts. Economic historians, aware of the impact that American arms had upon British technical observers at the 1851 London Crystal Palace Exhibition, have sought to explain in economic terms the divergent approaches to manufacturing in America and Great Britain. The general view has been that British manufactures remained labor intensive while American manufactures, more thoroughly mechanized, became capital intensive. The scholarly problem has been to explain why this was so. Business historians have done more than others to explain the eventual domination by American firms of international high-technology markets. A small but growing number of scholarly works have dealt with the fundamental changes in marketing techniques and the development of modern integrated manufacturing firms into "multinationals."[2]

Until the several strands of history leading up to America's worldwide industrial autonomy in the years between the two world wars are woven into a connected narrative, the catalytic effect of the American system of manufactures will not be generally recognized. Textbooks of American history will continue to credit Eli Whitney with interchangeable parts or with the invention of mass production, while Henry Ford will remain the last of a long line of "tinkerers"—one whose particular contribution was "to adapt to car manufacturing the production techniques first developed by small arms manufacturers."[3] Although American history textbook writers are aware of American prowess in manufacturing and marketing, their understanding of its development is sketchy and insecure. For that reason, generations of American students have learned a smattering of facts that seem unrelated and insignificant.

The American system of manufactures, seen as the first link of a chain leading to the technological invasion of cultures around the world, was a turning point of awesome implications. In the first half of the nineteenth century, in America, a change occurred in manufacturing methods that within another hundred years would alter fundamentally the quality of life for countless individuals in all countries. The first step was to mechanize production; almost immediately problems arose of selling all that could be produced. Ultimately, a new kind of industrial manager orchestrated the entire manufacturing and distribution system, from the procurement of raw materials to the delivery of the finished product at the appropriate moment to the ultimate consumer, wherever he or she may have been. The army whose banner was the American system of manufactures included entrepreneurs, who built and organized manufacturing plants in the United States; expatriate technicians and drummers, who made and sold the products abroad; and U.S. consuls and trade commissioners, who sent back technical and commercial information. Thousands of Americans, at home and abroad, found excitement and a sense of mission in extending American bounty to the rest of the world. Wherever the Americans were successful, whole cultures had to adjust to the new elements brought in by the eager crusaders.

This global dimension of the American system of manufactures emerges when the scholarly work already published is put together. It is the purpose of this chapter to review the record and call attention to its interrelatedness. There are five parts to my analysis. I shall first enter the shops and try to bring some order to the issue of interchangeable parts. Next, the factors, economic and otherwise, that help to explain the American system's rise will be reviewed. Thirdly, the rise of industrial shop management will be related to the modern integrated industrial corporation. Next, the process by which the American system was spread over the world will be explored. Finally, a cursory glance at the social effects of the American system may suggest how powerful it has been as an agent for social change.

Interchangeable Parts

Many if not most students of the American system of manufactures have assumed that the mechanics and entrepreneurs who developed it were bent primarily on attaining interchangeability of parts in their products in order to increase production and reduce costs. This notion is implicit in many articles and books, and Robert Howard found recently that twenty of twenty-two "scholars, arms makers, curators and students" agreed that "there is a general notion that interchangeable parts contribute to the lowest possible manufacturing costs."[4]

Before the Civil War, it is doubtful that anybody expected to save money through the use of interchangeable parts in military arms. John Hall, working on his interchangeable rifle in the 1820s, certainly understood the implications of interchangeability but he could never command a market large enough to show savings over conventional methods of turning, filing, and fitting.[5] The army's goal of interchangeability was based on the presumed advantages of being able to repair small arms in the field without requiring the skilled work of a gunsmith. Claims of cost savings for an undeveloped system were simply part of the rhetoric of asking Congress for large sums of money to support extensive mechanization.[6]

Howard has shown that pistol and revolver makers, including Samuel Colt, never produced fully interchangeable weapons, and that the making of a gas-tight, accurate revolver, even in the twentieth century, involves hand fitting of crucial parts. Howard joins Merritt Roe Smith in reaching the sensible conclusion that, in the period before the Civil War, complete interchangeability in a mechanism of considerable precision was not an economic goal but an "expensive luxury."[7]

Despite the necessity for hand fitting during assembly of their weapons, it must be emphasized that Colt and other makers of handguns mechanized the production of component parts to a very considerable degree. The Colt armory in Hartford was, in fact, a showplace of mechanization. Many different special-purpose machines, some of which served the armory for a hundred years, were designed by Elisha K. Root, and an original approach to drop forging by Root soon found wide application in many other industries.[8] Most of the parts of a Colt revolver were interchangeable, but the hand fitting in its final assembly was inevitable, given the practical impossibility of making at any reasonable cost parts that would fit very snugly and at the same time be interchangeable.

Claims that Eli Whitney had, before the War of 1812, produced muskets with interchangeable parts were considerably weakened by Robert Woodbury's delightfully titled article on "The Legend of Eli Whitney."[9] The added claim that Whitney was a pioneer of mechanization, having built a milling machine before 1820, was effectively dismissed by Edwin Battison and Merritt Roe Smith in a series of articles based upon new empirical evidence.[10]

Interchangeable parts and mechanized production are, in fact, two separate issues. To define the American system of manufactures as one involving concepts of interchangeable parts, as Woodbury did, is to imply that the advantages of mechanization had to wait until precise machine tools, precise gauges, and "certain techniques of mechanical drawing" were available.[11] Quite the contrary. The very large incremental cost advantages were (and are) to be gained in the coarser parts of the work.[12]

The primary money-saving feature of the American system of manufactures as it was observed by British visitors to the United States in the 1850s was the mechanization of turning, sawing, mortising, and other operations, both on wood and metal products. The Britons were enchanted by the Blanchard gunstock duplicating machine, which seemed to offer such obvious advantages as compared to manual shaping of gunstocks.[13] The application of interchangeability outside the armories depended upon the required precision of the product being made and upon the number of units that might be sold. Clocks, requiring a very low order of precision, were made of interchangeable parts as soon as the market could support the production of a few thousand a year.[14] Singer sewing machines, requiring nicely fitted bobbin assemblies, were not made with fully interchangeable parts until production reached several hundred thousand a year.[15] Textiles, too, were made in America in ways qualitatively different from British methods. David Jeremy, assessing American innovations during the early nineteenth century, pointed out that high operating speeds, greater mechanization, automatic fault detection, sequential movement of materials, and maximum use of inanimate power were characteristic features of textile manufacturing in New England.

Jeremy reminded us that "as components of the American system of manufacturing, these principles would later be absorbed into the concept and practice of early 20th-century mass production."[16] Students of the American system have generally ignored the textile industry as a leader in mechanization and automatic control of manufacturing processes, yet even this industry, so clearly shaped by its British origins, was responding to its American environment in ways that would eventually shake the world.

The first half of a definition of the American system of manufactures must, I believe, embrace the textile industry and the production of carriages, butter tubs, pulley blocks, plows, and lead pencils as well as small arms, sewing machines, and bicycles. In all of these industries, the trend was toward mechanized production of component parts, using special-purpose machines in a carefully planned sequence of operations. Not until the twentieth century did interchangeable parts become a generally viable component of the American system, and their presence or absence became crucial only in large mass-producing firms. The other half of the definition, to be dealt with a little later, will have to do with integrating the management of manufacturing and marketing.

Economic and Social Explanations

Since the publication in 1962 of H. J. Habakkuk's book comparing American and British technology, the central question regarding the American system has been why manufacturing technology in the United States developed along lines different from those in Great Britain.[17] The general assumption, reinforced by Habakkuk's arguments, was that American manufacturing was capital intensive and that American machines in the factory replaced skilled craftsmen. The problem was simply to show how economic forces had brought about the divergent technological styles in the two countries. Habakkuk concluded that the shortage and resulting high cost of skilled labor in the United States were reasons enough for the observed differences between American and British methods.

While some economic historians argued with Habakkuk on his own terms,[18] Nathan Rosenberg proceeded in a series of systematically related articles, now published as *Perspectives on Technology*, to move the whole argument to a different and vastly more fruitful level.[19] Rosenberg did what very few economic historians have done: he paid attention to the nature of technological change as well as its causes and results. In one article, he and Edward Ames point out the significance of qualitative differences between American and British products and machines—differences that were ignored by other economic historians.[20] In another article, Rosenberg suggests the difficulties that economists labor under by assuming that technology is quantitative and homogeneous and responds automatically to various economic stimuli. Rosenberg's examples of situations causing technological change that are not amenable to theoretical models involved production bottlenecks, recalcitrant labor, disruption of materials supply, and accidental mechanical failures. He has stressed throughout his work the diversity and complexity of technological change, even though he apparently believes that ultimate explanations must somehow be couched in economic terms.

Paul Uselding, in his recent painstaking review of "Studies of Technology in Economic History," pointed out that neoclassical theory, on which Habakkuk based his arguments, is particularly inadequate when analyzing qualitative differences in technology. He sums up Rosenberg's as well his own insistence upon historical rather than a priori reasoning in economic history with the observation that "historical reality cannot be pulled rabbitlike from some theoretical hat." [21]

Uselding notes in passing that the empirical work of Battison, Smith, and others who have studied antebellum American armories has supplied economic historians with most of what they know about the detailed historical aspects of the American system of manufactures. [22] Not only have the armories provided the grist for many historical mills, but the making of reapers, sewing machines, and bicycles has been treated merely as an expansion of the methods developed in the armories. The first empirical study of the three industries just named, made by David Hounshell in 1977, has shown unexpected qualitative differences that invalidate extrapolated data and in fact show that the competitive success of McCormick and Singer in their respective fields was *not* due to progressive technology. [23]

The economic importance of the capital-intensive American methods and the notion that machines were built primarily to eliminate workers' skills have been further undermined in empirical studies by Polly Earl and Edward Duggan. Earl has inquired into the extensive use before the 1870s of handpowered machines in woodworking. A representative example is the mortising machine, which has a guided chisel driven by a kick lever. This machine, with its wooden frame, was both simple and inexpensive. A joiner and a blacksmith could turn one out in a day's time, yet the mortising machine increased the productivity of its skilled operator by a factor of twenty or more. [24] This was but one of many woodworking machines designed not to eliminate but to speed up the work of skilled craftsmen. John Richards, writing in the 1880s, when steampowered machines were widely used, was familiar with both American and British woodworking practice. He saw the woodworker's job in a shop as a race between brains and muscle. The new machines had eliminated much of the muscular work, and the main business of the workman was to guide, direct, and take care of them. The muscular work was gone; the brain work remained. Richards noted that "in the great race for automatic machinery," manufacturers had incorporated power feeds in many of their machines despite the fact that a skilled workman's hand feed was better "both in quality and cost."[25]

Edward Duggan, in a study of the carriage and wagon industry in late-nineteenth-century Cincinnati, found that innovations were directed toward minimizing production time rather than eliminating skilled labor. Although contemporary observers found the carriage and wagon industry to be especially advanced in the use of production machinery, Duggan has noted that the proportion of capital devoted to machinery was only six percent.[26] Alfred Chandler has confirmed, in another context, that several of the capital-intensive industries of the 1880s were not expensive to enter.[27]

If a particular virtue of economic history is to eliminate the impressionistic approach that bedevils ordinary history, as some of its advocates claim, then it is only fair to insist that the implications of a term such as "capital intensive" in a particular situation be based upon empirical evidence. I suspect that the term brings to the minds of many economic historians, as it does to mine, a picture of a blast furnace or the assembly line at Henry Ford's River Rouge plant.

More significant in explaining Anglo-American differences than correcting the arithmetic of econometricians is the large number of instances in which decisions and trends appear to rest more on attitudes and enthusiasms than upon coolly calculated economic advantage.

John Sawyer's well-known 1954 article on the social basis of the American system of manufactures has not been superseded. He argued that the highly mechanized, highly productive style of manufacturing observed by British visitors both in the 1850s and the 1950s was based more on social assumptions than on economic stimuli. The ready adoption of new, more productive machines in factories, noted by Joseph Whitworth in 1853, was echoed in the report of a British industrial visitor after World War II, who thought that the "acceptance of the need for high productivity as an essential factor in industrial life is universal in American life."[28] Nathan Rosenberg's most recent judgment on the rise of the American system includes as a "major reason" the differences between British and Americans "in taste and demand, which in turn were deeply rooted in the social structure and distribution of income in each country."[29] George Daniels, in his provocative article on the "big questions" in American technology, asserted the importance of national goals and preferences in bringing the

American system into being.[30] John Burke, in responding to Daniels's paper, reminded us that innovation is not the product of a society as a whole but of a small minority, which must then convince the public that the new innovations enhance commonly held social values.[31]

Merritt Roe Smith, in his prize-winning Harpers Ferry Armory study, developed the idea that agencies of social control, particularly schools and churches, have an important role in defining common values and goals and instilling "a conviction that such norms are both necessary and good."[32] The family, too, is an agent of social control, and during the nineteenth century some surprising notions were given to children along with their mothers' milk. We are likely to think of the movement for efficiency in the home as growing out of the industrial-efficiency movement of the early twentieth century. An inspection of American homemaker's manuals, published since the 1830s, suggests that the process may almost have been the reverse, and that an efficiency-ridden populace was waiting to march onto the twentieth-century assembly lines. Catharine Beecher, in her *Treatise on Domestic Economy* first published in 1841, reminded her readers that "Christianity teaches, that, for all the time afforded us, we must account to God; and that we have no right to waste a single hour." In her later and better-known *The American Women's Home* (1869), written with the help of her sister, Harriet Beecher Stowe, the reader learns that a well-planned American cottage economizes "time, labor, and expense."[33] In the 1930s, in the midst of the Great Depression, Robert and Helen Lynd found in Middletown that "progress is the law of life" and that progress can be promoted by individuals who are enterprising, practical, and efficient.[34]

Another ideological thread that encouraged the rise of the American system of manufactures appeared first in England in 1832, when Andrew Ure wrote, "The most perfect manufacture is that which dispenses entirely with manual labour." Ure thought that Adam Smith's concept of division of labor was quite obsolete, and that "the constant aim and tendency of every improvement in machinery [is] to supersede human labour altogether, or to diminish its cost, by substituting the industry of women and children for that of men; or that of ordinary labourers, for trained artisans."[35] Twenty years later, when Samuel Colt testified on his pistol making before a select committee of the Parliament, he boasted that "there is nothing that cannot be produced by machinery."[36] Charles Fitch, whose long "Report on the Manufactures of Interchangeable Mechanism" was published in the 1880 census reports,[37] wrote a short article that appeared in 1884 in the *Magazine of American History*. Entitled "Rise of a Mechanical Ideal," the article exulted in the idealism of the "mechanics" who developed the interchangeable system of manufactures.[38] In his census report, Fitch asserted that "in some respects, this system has been one of the chief influences in the rapid increase in the national wealth." [39]

Just before World War I, Lloyd R. Smith, son of the founder of the A. O. Smith Corporation in Milwaukee, maker of automobile frames, set out to build "a plant to run without men." Eight million dollars and ten years later his plant was running, turning out an automobile frame every ten seconds. It was an engineering tour-de-force, starting with strip steel, mechanically forming and assembling the frame elements, and automatically placing and heading a million rivets a day. Men were used only to monitor the automatic operations. Yet Smith wrote lugubriously in the *Magazine of Business* about his "failure" to attain his goal, having reduced the need for men only to twelve man-minutes per frame. It was obvious from the article, in which Smith called himself a "business radical," that the incentive to build the fully automatic plant was a psychological one. Neither demand for frames nor scarcity of labor was calling for automatic machinery. The impetus was clearly an obsession carried to a fanatical extreme to dispense with labor.[40]

Just after World War II, two young Canadians published an article in *Fortune* entitled "Machines Without Men." The rhetoric was up to date, speaking of an "automatic, scientific, flexible, and functional" factory, with machines specialized in terms of their functions rather than in terms of their products. The elements to build the automatic factory were known, the authors asserted, but a new philosophy was needed to make it possible for them to put the elements together in a new way.[41] The "mechanical ideal" has been around for a hundred and fifty years, and it is alive and well in the engineering press, where each new generation of engineers reinvents the wheel.

The "New Factory System"

The rise of scientific management in manufacturing shops and of new strategies in business and marketing contributed to the effectiveness of the American system of manufactures and enabled it to capture new markets for its products.

Going beyond Taylorism and looking at industrial shop administration as a whole, Daniel Nelson has filled in a number of conspicuous lacunae in traditional accounts of shop management. In so doing, he has brought a fresh view to bear on the significance of foremen in the American system of manufactures. Before 1880, in some advanced New England shops, foremen acted as contractors who agreed to furnish to the shop manager a certain number of parts or subassemblies in a given time. The foreman was responsible for hiring and paying his own help. This system attracted a number of talented and original mechanics; the principals of the machine-tool firms of Pratt & Whitney and of Warner & Swasey, for example, had earlier been contractor-foremen.[42]

Between 1880 and 1920, radical changes occurred in the relationships between workers and industrial management. Efforts to control costs, to increase the productivity of the worker, and to transfer all initiative from the shop floor to an arm of management above the level of foreman, were varied in design but constant in exerting pressure for changes that would benefit management. Frederick W. Taylor's system tried to usurp management's power of decision by interposing a planning department, to be directed by engineers, between the levels of shop foremen and business management. Shop foremen were then emasculated by the shift to a system of "functional foremen," each responsible for a part of the traditional foreman's job, such as "gang boss," "inspector," "repair boss," and "disciplinarian."[43] Taylor's call for a "complete mental revolution," both by management and worker, in order to have a fully cooperative work force, went unheeded;[44] workers understood very well what Taylor expected of them, and management was not about to relinquish control to a soviet of engineers in the planning department.

Bits and pieces of Taylor's system were adopted by businessmen and important savings in production costs were realized through attention to what went on in the shops, yet by the time Taylor came on the scene production problems were no longer central in manufacturing firms.[45] The capability already existed to make more units than could be sold in the absence of a well-planned strategy of marketing.

A marketing revolution occurred in the latter half of the nineteenth century as mass markets were tapped by department stores, chain stores, and mail-order houses. These buying and selling organizations, availing themselves of the telegraph, railroads, steamships, and improved postal service, coordinated the flow of goods from many manufacturers to many customers. In some of the high-volume mechanical industries, such as those making sewing machines and agricultural implements, the producing firm took over marketing responsibilities and built dealer organizations that were the prototypes of modern integrated marketing operations.[46]

Charles Fitch, in the 1880 census report, observed that "the economy of large manufacture and uniform methods is largely due to the fact that so little labor is wasted."[47] The absence of waste depended, however, on careful coordination of the flow of supplies and finished products as well as careful shop management. As Alfred Chandler has pointed out, pressures grew in large firms to administer every step of the sequence from extracting raw materials to delivering finished products to the customer's doorstep. Modern factory organization grew from the procedures of progressive firms that adopted and developed the American system of manufactures in the nineteenth century.[48]

Missionary Zeal

Just as continued improvements in manufacturing processes depended as much on ideological push as on economic pull, so the spread of American products to the ends of the earth called for salesmen and entrepreneurs with missionary zeal to seize economic opportunities in other countries. Even in the late eighteenth century, enterprising Americans had journeyed to England to seek support for their inventions and to take advantage of an organized commercial network, which had not yet been developed in the United States.[49] In the 1840s, the flow of British-built railway locomotives to the United States was reversed, and Philadelphia-built Norris locomotives were sent to England. Two of the Norris brothers journeyed to Austria to build locomotives for the emperor, and other Americans spent years in Russia building railroads and rolling stock for the czar.[50] By the time the American system of manufactures assumed recognizable form, an American tradition of seeking opportunities wherever they existed was well established.

Following the American triumph at the London Crystal Palace Exhibition in 1851, several American entrepreneurs, including Samuel Colt, Day and Newell, and Richard Hoe, either established manufacturing shops in England or had their products built in English machine shops.[51] When American machine tools were purchased for the new Enfield Armory, near London, James Henry Burton of Virginia accepted a five-year appointment as chief engineer of the armory; he was to supervise the installation and operation of the American machinery. In the course of his work at Enfield, he came to know Thomas Greenwood, an English machine-tool builder who supplied tools from American patterns to supplement those imported from the United States. In 1871, Greenwood arranged to install in Russia a small-arms factory patterned after the American system, and in 1872 Burton took a team of English workmen to Tula, ninety miles south of Moscow, and spent nearly two years in building the new works and in instructing the Russian supervisors and workmen in the intricacies of the American system of manufactures.[52]

By the end of the nineteenth century, American products were flowing to many points on the globe. Singer sewing-machine factories had been operating in Austria and Scotland since the 1880s; a new factory was opened in Russia in 1902 and another in Prussia in 1904.[53] The first foreign Eastman Kodak factory was established in England in 1889, and a marketing network was established that extended through Europe to Asia, Africa, and Australia.[54]

Early in the twentieth century, there was a commonplace notion in the United States that the only American manufactured products competitive in world markets were those "in which cheapness and not superior quality has been the determining factor." To refute this charge, Consul-General Frank Mason, in Paris, published in the *Monthly Consular and Trade Reports* a list of American products that had "won their way in European markets by sheer force of superior quality." These included machine tools, typewriters, sewing machines, agricultural implements, fine hardware such as locks and hinges, shoes, silverware, and watches. He noted that these goods belonged to "the class of products turned out in large quantities, mainly by the use of ingenious and highly effective machinery... operated by... a multitude of automatic workmen." He went on to deplore the absence of artistic, handmade American products in the French market, and he thought that the French could teach Americans a great deal in that department; but he was clear and convincing in his praise of products made according to the American system of manufactures.[55]

By the end of World War I the "new factory system," embodying a scheme of scientific management or one of the alternative systems, had become a permanent and central part of the industrial scene in the United States.[56] A campaign to transfer Taylorism to Europe developed in the 1920s, as American and French advocates carried on the crusading spirit of Taylor and his followers.[57] A report published in 1927 by the International Labor Office in Geneva hailed the "blossoming" of the scientific management movement after the war. The author conceded that a number of psychological problems had yet to be solved, but he thought that "co-ordinated action in the matter of research, information, and education" would convert the unbelievers and bring to Europe the undoubted blessings of this American scheme. "As regards individualism, which is opposed to the standardisation of products, it is surely possible to educate consumers and to show them where their true interests lie. The Americans have done this."[58]

In Europe, as in the United States, management systems were modified as individual firms adopted some features while rejecting others. On the other hand, American ideas were congenial to many European managers, and the undoubted success of systematic factory management in America forced the issue in Europe. Charles Maier has shown how European traditions and social currents of the 1920s were favorable to the transfer of American methods. In Germany, Fordism was more attractive than Taylorism because Taylorism was concerned only with the management of labor, while Fordism stressed the reorganization of the entire production process.[59]

In summing up the overseas campaign that commenced at the beginning of the nineteenth century and assumed a driving messianic character after the 1851 Crystal Palace Exhibition, it is clear that American consumer products, machine tools, and management methods became imbedded in the social fabric of countries all over the world.

Social Benefits and Costs

The social benefits and costs of the American system of manufactures, when measured on a worldwide scale, remain largely unexplored. American authors have not yet tried to assess the social effects of the Americanization of certain aspects of foreign cultures. A few books, responding to particular pressures exerted by the American campaign, have appeared from time to time. In 1901 and 1902, for example, three books were published in England responding (as reflected in the title of one) to *The American Invasion*.[60] Many short articles have been published in trade journals, either welcoming or decrying the latest American initiatives, but nobody to my knowledge has studied this literature systematically.

One of the central issues in assessing the social results of the American system is its effect upon workers in the factory. Fifty years ago, Henry Ford formulated his view of the question. In the article on "mass production" in the thirteenth edition of *Encyclopaedia Britannica*, published in 1926 over Henry Ford's name, one finds this passage: "It has been debated whether there is less or more skill as a consequence of mass production. The present writer's opinion is that there is more. The common work of the world has always been done by unskilled labor, but the common work of the world in modern times is not as common as it was formerly." In disposing of the charge that assembly-line work was monotonous, Ford asserted that "monotony does not exist as much in the shops as in the minds of theorists and bookish reformers."[61] The ensuing debate has been carried on pretty much at this offhand level, with differing opinions endlessly stated and restated.

A very useful contribution to the discussion was made in 1974 by the late Harry Braverman in his book *Labor and Monopoly Capital*. His analysis is based upon a sure knowledge of the technical processes from design through assembly of the finished product, and his distinction between mental work and manual work is a valid one. He recognized that much of the revolutionary power of scientific management rested upon its separation of mental work from manual work. Once this separation of conception from execution has been made, the two aspects of the labor process can be carried out in separate locations by separate bodies of workers. The mental work involved in planning manufacturing processes was appropriated by engineers, who decide not only what work will be done in the shop but also in minute detail how it must be done. As planning has been refined and systematized, the need for mental work in the shop has been reduced to a minimum. Despite the much greater sophistication of machinery in modern shops, the need has increased for docile labor, willing to accept without question the prescribed and circumscribed jobs that have been devised by mental workers who have only a vicarious understanding of manual work.[62] The debate will continue, but any serious discussion of the issue will have to take into account Braverman's analysis.

Conclusion

This review has been guided by my reading of published works that deal with the American system of manufactures and by a view that historians of technology, business, and economics should be able to supply textbook writers, journalists, and the general public with a plausible, interesting, and relevant picture of a movement that in less than two hundred years has profoundly changed the conditions of life on this planet. While we take our "big story" to a wider audience, we also need to look more carefully at the technical details of production in manufacturing plants, particularly during the period when quantities increased dramatically and the solutions to problems of finishing and assembly required heroic measures both in the shop and in the business office. Business and economic historians can use more synthetic works with informed technical discussions set in a matrix of economic and social reality, as in Merritt Roe Smith's and Reese Jenkins's recent books. Historians of technology must take seriously the relevance to their technical and social concerns of business decisions and economic systems, even while increasing their own understanding of the technical imperatives that obsess innovative technologists.

In conclusion, I offer a definition of the American system of manufactures, based on its historical development, that I hope will underscore its scope and influence. The American system of manufactures grew from a demonstration and conviction that large numbers of uniform products could be produced most satisfactorily with the help of machines, subdividing the operations of manufacturing and employing special-purpose machines wherever possible. In its economically most promising form, the system appeared first in the Portsmouth Dockyards in England and in the wooden-clock factories of New England.[63] To extend the principle of quantity production to metal mechanisms of great precision, a forcing process within American armories, financed by the Army, was carried out in the first half of the nineteenth century. Disposing of the great quantities of sewing machines and bicycles that could be produced under the American system required the development of new marketing strategies. These encouraged, in turn, new structures of business organization. Engineers who were drawn to management problems in the 1880s developed systematic or scientific management schemes that dealt with planning and administration within manufacturing shops. The larger problems of integrating supply, manufacturing, and marketing functions were attacked in a series of consolidations in the 1880s and 1890s. Finally, the migration of American products and American methods to other countries and eventually throughout the world depended upon salesmen and technicians who formed, in the words of a Singer official, "a living moving army of irresistible power, peacefully working to conquer the world."[64]

The saga of this American conquest, essentially completed in the decade after World War I, remains to be told in such a way that it will take its place in textbooks (which supply a nation's view of itself) alongside other great historical movements that have changed forever the conditions under which human beings live.

Notes

1. Nathan Rosenberg, *The American System of Manufactures* (Edinburgh, 1969), p. 5. Rosenberg says the term originated in London in 1851, in response to the American products exhibit in the Crystal Palace Exhibition. The exact point of origin is elusive, however, as is the first use of the term by historians.

2. Glenn Porter and Harold C. Livesay, *Merchants and Manufacturers: Studies in the Changing Structure of Nineteenth-Century Marketing* (Baltimore, 1971); Alfred D. Chandler, *The Visible Hand: The Managerial Revolution in American Business* (Cambridge, Mass., 1977); Mira Wilkins, *The Emergence of Multinational Enterprise* (Cambridge, Mass., 1970).

3. Irwin Unger, *American History II (Reconstruction to Present)*, Monarch College Outlines (New York, 1971), p. 134. This "cram" text is designed to enable a student using any college American history survey to acquire a distillation of essential facts and ideas. Bernard A. Weisberger, *The Impact of Our Past: A History of the United States* (New York, 1972), p. 317, has Whitney making rifles with interchangeable parts before the War of 1812. "Inspectors for his biggest customer, the United States government, were amazed when workmen put together finished guns from piles of barrels, locks,

triggers, and guards. Later the first assembly lines were set up." Weisberger's is a current high-school textbook from a historically oriented publisher.

4. Robert A. Howard, "Interchangeable Parts Reexamined: The Private Sector of the American Arms Industry on the Eve of the Civil War," *Technology and Culture* 19 (1978): 633–49; quotations on pp. 635, 636. Paul Uselding, "Elisha K. Root, Forging, and the 'American System'," *Technology and Culture* 15 (1974): 543–68. Uselding notes (p. 548) that the full economic advantage of interchangeable manufacture is not realized until production is large enough to capture economies of scale. Chandler, *The Visible Hand* (note 2), p. 281, believes that economies of speed are more important than economies of scale.

5. Merritt Roe Smith, *Harpers Ferry Armory and the New Technology* (Ithaca, N.Y., 1977), pp. 191–218.

6. Ibid., p. 325.

7. Howard, "Interchangeable Parts" (note 4), p. 649.

8. Uselding, "Elisha K. Root" (note 4). Uselding's theory of diminishing returns that encouraged a shift from milling to drop forging, based as it is upon conjecture, does not appeal to me as reasonable.

9. Robert S. Woodbury, "The Legend of Eli Whitney and Interchangeable Parts," *Technology and Culture* 1 (1960):235–53.

10. Edwin A. Battison, "Eli Whitney and the Milling Machine," *Smithsonian Journal of History* 1 (Summer 1966): 9–34; Battison, "A New Look at the Whitney Milling Machine," *Technology and Culture* 14 (1973): 592–98; Merritt Roe Smith, "John Hall, Simeon North, and the Milling Machine: The Nature of Innovation Among Antebellum Arms Makers," *Technology and Culture* 14 (1973): 573–91. For a connected account of New England experience, see Felicia J. Deyrup, *Arms Makers of the Connecticut Valley: A Regional Study of the Economic Development of the Small Arms Industry, 1790–1870*, Smith College Studies in History 33 (Northampton, Mass., 1948). For additional facts and insights regarding the mechanization of the Springfield Armory between 1795 and 1833, emphasizing the pitfalls of assuming that managers made economically rational decisions and policies, see Paul Uselding, "An Early Chapter in the Evolution of American Industrial Management," in *Business Enterprise and Economic Change*, ed. Louis P. Cain and Paul Uselding (Kent, Ohio, 1973), pp. 51–84.

11. Woodbury, "The Legend of Eli Whitney" (note 9), p. 247.

12. Furthermore, the horseshoe- and railroad-spike-forging machines of Henry Burden in Troy, N.Y., were not cost-saving improvements of existing methods of producing horseshoes and railroad spikes; rather they were new and original machines that made possible a quantum jump in production at a marked cost saving. They are, I believe, an integral part of the American system. See Paul Uselding, "Henry Burden and Question of Anglo-American Technological Transfer in the Nineteenth Century," *Journal of Economic History* 30 (1970):312–37.

13. Rosenberg, *The American System of Manufactures* (note 1), pp. 49, 137.

David A. Hounshell, "From the American System to Mass Production: The Development of Manufacturing Technology in the United States, 1850–1920" (Ph.D. diss., University of Delaware, 1978). Hounshell points out (pp. 380–81) that what impressed British visitors about manufacturing methods "was the 'application of special tools to minute purposes.' *That* was what the British meant when they thought of the American system of manufactures, and that was what was almost universal in all kinds of manufacturing establishments in the United States."

14. Howard, "Interchangeable Parts" (note 4), p. 633, notes the ease with which interchangeability could be attained in the clock industry. John J. Murphy, "The Establishment of the American Clock Industry: A Study in Entrepreneurial History" (Ph.D. diss., Yale University, 1961), pp. 56–57, sees the early clock industry as the first link in the chain that was to become the American system of manufactures. John J. Murphy, "Entrepreneurship in the Establishment of the American Clock Industry," *Journal of Economic History* 26 (1966): 169–86, describes quantity production of clocks before 1814. It is surprising that nobody has built on Murphy's solid work.

15. Hounshell, "From the American System to Mass Production" (note 13), pp. 168–69. The Singer decision to insist on full interchangeability was made in 1883, when production was about 600,000 per year.

16. David J. Jeremy, "Innovation in American Textile Technology during the Early 19th Century," *Technology and Culture* 14 (1973): 40–76.

17. H. J. Habakkuk, *American and British Technology in the Nineteenth Century: The Search for Labour-Saving Inventions* (Cambridge, 1962). Reviews of this book were favorable although the absence of empirical data was noted by Rosenberg and Jacob Schmookler. The anonymous reviewer in the *Times Literary Supplement* (27 July 1962, p. 539) thought it "above all a lesson in care and precision of thought, rather like a powerful end-game study in chess." The results were familiar, even "jejune," but the book's purpose was not so much to discover as to connect. I have not attended end-game studies, but I have been put off by Habakkuk's admittedly "*a priori* arguments in the light of a small number of facts" (p. 92). Excerpts of this book, comprising the arguments regarding the American system (pp. 11–34, 43–63, 83–90) have been reprinted in *Technological Change: The United States and Britain in the Nineteenth Century*, ed. S. B. Saul (London, 1970), pp. 23–76.

18. E.g., Peter Temin, "Labor Scarcity and the Problem of American Industrial Efficiency in the 1850's," *Journal of Economic History* 26 (1966): 277–98. Termin notes (p. 287) that it is impossible to admit that a businessman might behave in an economically irrational manner "because it denies the usefulness of [economic] reasoning."

19. Nathan Rosenberg, *Perspectives on Technology* (Cambridge, 1976). This volume will richly repay careful reading. Rosenberg's seminal chapter (1) on machine tools states his principle of "technological convergence," which helps to explain the diffusion of ideas. He tells us (p. 83) that "technological change is an extremely complicated social process . . . having dimensions that do not fall conveniently within the boundaries of any single academic discipline." He notes (p. 102) that economists have been preoccupied with "economic activity

within the firm as the sole source of productivity improvements." He has sorted out many, many erroneous or questionable assumptions of economic historians regarding technology. It would be useful to have a similarly constructive work of criticism in technological history.

20. Edward Ames and Nathan Rosenberg, "The Enfield Armory in Theory and History," *Economic Journal* 78 (1968): 827–42.

21. Paul Uselding, "Studies of Technology in Economic History," in *Recent Development in the Survey of Business and Economic History: Essays in Memory of Herman E. Kroos*, ed. Robert Gallman (Greenwich, Conn., 1977), pp. 159–219; quotation on p. 165. Uselding notes (p. 170) the importance of qualitative changes in technology. He cites (p. 178) an article by R. A. Church on the shift of leadership in watchmaking from Britain to America and Switzerland. The reasons were not so much factor substitution as a qualitatively different array of watchmaking machinery, and British watchmakers apparently were not interested in a low-price market. Hounshell, "From the American System to Mass Production" (note 13), pp. 316–19, notes the appearance in the Midwest of punch presses to produce sheet-metal stampings for bicycles, making parts qualitatively different from those forged or hogged out of solid metal as in New England armory practice.

22. Uselding, "Studies of Technology in Economic History" (note 21), pp. 173–74.

23. Hounshell, "From the American System to Mass Production" (note 13), pp. 180–81, 382–83.

24. Polly Ann Earl, "Craftsmen and Machines: The Nineteenth Century Furniture Industry," in *Technological Innovation and the Decorative Arts*, ed. Ian M.G. Quimby and Polly Ann Earl (Charlottesville, Va., 1974), pp. 307–29. The estimates of time required to build a mortising machine and its increase of productivity are mine. A perusal of the illustrations in the *Scientific American*'s "American Industries" series, starting in 1879, will convince an observer that a significant proportion of production machinery in mass-production industries was simply designed and lightly built. See, for a list of the articles, *Technology and Culture* 17 (1976): 82–92. David Pye, *The Nature and Art of Workmanship* (Cambridge, 1968), makes a useful distinction between "workmanship of risk" and "workmanship of certainty." The unsophisticated machines under discussion reduced drastically the "workmanship of risk" and thus increased speed of execution by permitting jigged or guided execution of machine-aided hand processes. George Daniels, "The Big Questions in the History of American Technology," *Technology and Culture* 11 (1970): 1–35, suggests (p.16) that the presence of skills, not their absence, may have been an important factor in the rise of the American system.

25. John Richards, *On the Arrangement, Care, and Operation of Wood-Working Factories and Machinery* (London, 1885), p. 119.

26. Edward P. Duggan, "Machines, Markets, and Labor: The Carriage and Wagon Industry in Late-Nineteenth-Century Cincinnati," *Business History Review* 51 (1977): 308–25, esp. pp. 309, 315. The total capital was distributed as follows: machinery, 6%; buildings, 6%; land and rent, 26%; raw materials, 11%; in-process and finished goods on hand, 14%; trade capital, cash,

Ferguson

receivables, 37% of capital. In foundries and machine shops, machinery accounted for 31% of capital.

27. Chandler, *The Visible Hand* (note 2), p. 249.

28. John E. Sawyer, "The Social Basis of the American System of Manufacturing," *Journal of Economic History* 14 (1954): 361–79; quotation on p. 364.

29. Rosenberg, *Perspectives on Technology* (note 19), p. 287. Rosenberg quotes Alfred Marshall (p. 105): "... the development of new activities giving rise to new wants, rather than of new wants giving rise to new activities." This restates Thorstein Veblen's "... always and everywhere, invention is the mother of necessity" (in his *The Instinct of Workmanship* [1914; reprinted ed., New York, 1964], p. 314). Russell I. Fries, "British Response to the American System: The Case of the Small-Arms Industry after 1850," *Technology and Culture* 16 (1975): 377–403, observes that "receptivity to change is an important variable" in the "complex process" of technological change (p. 403). Even Peter Temin, in his *Causal Factors in American Economic Growth in the Nineteenth Century* (London, 1975), p. 32, avers that "a distinctive American style in manufacturing methods began to appear."

30. Daniels, "The Big Questions" (note 24), pp. 6, 21. John A. Kouwenhoven, "Waste Not, Have Not, A Clue to American Prosperity," *Harper's Magazine* 218 (Mar. 1959): 72–79, 81, in arguing that American technological prowess came first and a state of plenty followed, takes issue with David M. Potter, who in his *People of Plenty* (Chicago, 1954) assumed that abundant resources were the primary cause of plenty.

31. Daniels, "The Big Questions" (note 24), pp. 23–24.

32. Smith, *Harpers Ferry Armory* (note 5), p. 330n.

33. Catharine Beecher, *A Treatise on Domestic Economy* (New York, 1846), p. 181; Catharine Beecher and Harriet Beecher Stowe, *The American Woman's Home* (New York, 1869), p. 429.

34. Robert S. and Helen Merrell Lynd, *Middletown in Transition* (New York, 1937), pp. 405–6.

35. Richard Rosenbloom, "Men and Machines: Some 19th-Century Analyses of Mechanization," *Technology and Culture* 4 (1964): 489–511. This article reviews the responses to mechanization of Babbage, Ure, and Marx in England and Wells and Wright in America.

36. Great Britain, House of Commons, Sessional Papers, 1854, vol. 18, *Report from the Select Committee on Small Arms together with the proceedings of the committee* . . . 12 May 1854, p. 99.

37. Charles H. Fitch, "Report on the Manufactures of Interchangeable Mechanism," in *Report on the Manufactures of the United States at the Tenth Census* (Washington, D.C., 1883), 2: 611–704.

38. Charles H. Fitch, "Rise of a Mechanical Ideal," *Magazine of American History* 11 (1884): 516–27.

39. Fitch, "Report on the Manufactures of Interchangeable Mechanism" (note

37), p. 615.

40. L.R. Smith, "We Build a Plant to Run Without Men," *Magazine of Business* 55 (Feb. 1929): 135–38, 200. Two related pieces by Smith are in the next two issues: Mar. 1929, pp. 264–66; and Apr. 1929, pp. 398–99.

41. E.W. Leaver and J.J. Brown, "Machines Without Men," *Fortune* 34 (Nov. 1946): 165, 192, 194, 196, 199–200, 203–4; quotation on p. 165. Nineteenth-century machine tools—milling machines, lathes, drill presses—were specialized in terms of function rather than in terms of product. The irresistible urge to build machines to replace workers is illustrated in William B. Parsons, "Engineering Development of the Far East," *Engineering Magazine* 19 (July 1900): 481–92. Among the "most needed" things in China, after mechanical transport and hydraulic and mining machinery, were "those machines which can compete against a very low-priced manual labor."

42. Daniel Nelson, *Managers and Workers: Origins of the New Factory System in the United States 1880–1920* (Madison, Wis., 1975), pp. 36–38. Joseph W. Roe, *English and American Tool Builders* (New Haven, Conn., 1916), chap. 14, "The Colt Workmen," especially p. 178; see also p. 262.

43. Nelson, *Managers and Workers* (note 42), chaps. 3, 4. Hugh G. J. Aitken's *Taylorism at Watertown Arsenal: Scientific Management in Action 1908–1915* (Cambridge, Mass., 1960) is the most perceptive study of Taylor's aims and work.

44. Frederick Winslow Taylor, *Scientific Management, Comprising Shop Management, The Principles of Scientific Management, Testimony Before the Special House Committee* (New York, 1947), p. 250.

45. Alfred D. Chandler, "The Structure of American Industry in the Twentieth Century: A Historial Overview," *Business History Review* 43 (1966): 255–98. Problems of marketing, research and development, and integration of business functions were often more urgent than production problems. Duggan, "Machines, Markets, and Labor" (note 26), suggests quite accurately that managers of carriage and wagon works had more to worry about than production. Murphy, in his dissertation on the "American Clock Industry" (note 14), pp. 56–57, lists the method of distributing clocks as one of Eli Terry's several contributions to the mass production of clocks. Porter and Livesay, in *Merchants and Manufacturers* (note 2), p. 193, show that around 1900 the National Cash Register Company's problems were in mass distribution not in mass production. Glenn Porter, "Management," in *A History of Technology: The Twentieth Century*, ed. T. I. Williams (New York, 1978), pp. 77–92, avers that new organizational strategies of large firms were more significant than Taylorism.

46. Chandler, *The Visible Hand* (note 2), chap. 7. Hounshell, in "From the American System to Mass Production" (note 13), pp. 180–81, 382–83, found marketing to be more important than production in the success of the Singer and McCormick companies. He also noted (pp. 301, 327 n.45) annual model changes in bicycles in the early 1890s and the "loading" of bicycles with accessories.

47. Fitch, "Report on the Manufactures of Interchangeable Mechanism" (note

37), p. 713.

48. Chandler, *The Visible Hand* (note 2), pp. 281–83.

49. For example, Oliver Evans sent a model of his automatic flour mill to England with Robert Leslie in 1792, Amos Whittemore obtained an English patent for his textile-card machine, and George Clymer took his influential printing press to England in 1817. Greville and Dorothy Bathe, *Oliver Evans* (Philadelphia, 1936), p. 29; *Dictionary of American Biography*, s.v. Amos Whittemore and George Clymer. Jacob Perkins was an expatriate American in England whose system of "siderography" was used to produce British postage stamps for decades after 1839. Elizabeth M. Harris, "Experimental Graphic Processes in England 1800–1859," *Journal of the Printing Historical Society* 4 (1968): 33–86; see pp. 72–73.

50. P.C. Dewhurst, "The Norris Locomotive," Railway and Locomotive Historical Society, *Bulletin* 79 (1950): 1–80. William and Octavius Norris went to Austria in 1844. Joseph Harrison, Thomas Winans, and Andrew M. Eastwick went to Russia before 1845. Harrison returned in 1852, and the others remained to continue the work. *Dictionary of American Biography*, s.v. Joseph Harrison.

51. For the Colt pistols, see Howard L. Blackmore, "Colt's London Armoury," in *Technological Change: The United States and Britain in the Nineteenth Century*, ed. S.B. Saul (London, 1970), pp. 171–95. For the Day and Newell lock works, operated by Alfred C. Hobbs, see Rosenberg, *The American System of Manufactures* (note 1), p. 11. For the Hoe printing presses, built under license by Joseph Whitworth in Manchester, see A.E. Musson, "Joseph Whitworth and the Growth of Mass-Production Engineering," *Business History* 17 (1975): 109–49; reference on p. 128. Between 1865 and 1870 a large Hoe plant was established in London; see *Dictionary of American Biography*, s.v. Richard Marsh Hoe.

52. Edward C. Ezell, "James Henry Burton et le transfert du 'système américain' aux arsenaux du gouvernement impérial russe," *Le Musée d'Armes* (Liège) *Bulletin* 5 (1977): 13–20. (Delivered in English at the 20th Anniversary Conference of the Hagley Program, Eleutherian Mills-Hagley Foundation, Wilmington, Delaware.) A German traveler in Russia in the 1840s told of seeing at Ije, near Kasan, several hundred miles east of Moscow, an arms manufactory in which "the experiment has often been successfully made of taking to pieces a large number of muskets, and then, from the promiscuous heaps of similar parts, to put them all together again." Adolph Ermann, *Travels in Siberia*, trans. W.D. Cooley, 2 vols. (London, 1848), 1: 171–73. Gene S. Cesari, "American Arms-Making Machine Tool Development, 1798–1855" (Ph.D. diss., University of Pennsylvania, 1970), pp. 224–25, writes of Pratt & Whitney "packaged armories being sent abroad" to Germany and of Colt's contract to provide machinery for a Russian armory in 1856. Penrose Hoopes, in a book review in *Technology and Culture* 7 (1966): 96, writes: "The Soviet [watch] industry had its origin in the Deuber-Hampden Watch Company of Canton, Ohio, which was purchased by the Soviets in 1929, and moved bodily to Russia."

53. Robert B. Davies, *Peacefully Working to Conquer the World: The Singer*

Sewing Machine Company in Foreign Markets, 1854–1920 (New York, 1976). I learned these details in an informative review of the book by David Hounshell in *Technology and Culture* 18 (1977): 700–3. American-made parts were being assembled in Scotland as early as 1867.

54. Reese V. Jenkins, *Images and Enterprise: Technology and the American Photographic Industry 1839 to 1925* (Baltimore, 1975), pp. 117, 178.

55. U.S. Dept. of Commerce and Labor, Bureau of Manufactures, *Monthly Consular and Trade Reports*, No. 317 (Feb. 1907) [Doc.ser.no. 5176], pp. 60–61. A paragraph about each of the products listed enlarges on the general statements quoted. It is clear that quality not price was the determining factor in American success.

56. Nelson, *Managers and Workers* (note 42). "The New Factory System" is part of the subtitle of Nelson's book.

57. L. Urwick and E.F.L. Brech, *The Making of Scientific Management*, vol. 1: *Thirteen Pioneers* (London, 1951), p. 95, has this passage describing Henry le Chatelier (1850–1936): "His forceful and persistent elucidation of Taylor's philosophy, principles, and technique was a driving force that secured their acceptance in France and over large parts of continental Europe." Charles S. Maier, "Between Taylorism and Democracy: European Ideologies and the Vision of Industrial Productivity in the 1920s," *Journal of Contemporary History* 5 (1970): 27–61, notes the sponsorship by Edward Filene and his XXth Century Fund of international congresses to promote Taylorism (p.54).

58. Paul Devinat, *Scientific Management in Europe*, International Labour Office, Studies and Reports, Series B (Economic Conditions), No. 17 (Geneva, 1927), pp. 166–68.

59. Maier, "Between Taylorism and Democracy" (note 57).

60. Fred A. McKenzie, *The American Invaders* (London, 1901); B. H. Thwaite, *The American Invasion* (London, 1902); W.T. Stead, *The Americanization of the World* (London, 1902); cited originally in Wilkins, *The Emergence of Multinational Enterprise* (note 2), p.171, and noted in Chandler, *The Visible Hand* (note 2), p. 571 n. 31.

61. Henry Ford, "Mass Production," in *Encyclopaedia Britannica*, 13th ed. (1926), suppl. vol. 2, pp. 821–23. Nathan Rosenberg more than once has noted Ford's aphorism "In mass production there are no fitters." Its significance was finally impressed upon me by David Hounshell. Ford, in this article, writes: "The necessary precedent condition of mass production is a capacity, latent or developed, of *mass consumption*, the ability to absorb large production." The italics are Ford's. He also notes that "the greatest development of mass production methods has occurred in the production of conveniences [as opposed to necessities]." In 1914, the automobile was considered by bankers to be a pleasure vehicle and an extravagance; they were, therefore, unwilling to loan money to an automobile purchaser. A comprehensive survey of the aspects of automobiles of interest to a promoter was prepared by the advertising department of Curtis Publishing Co. *(Saturday Evening Post)*, Philadelphia, based on a 14,000-mile tour of the United States. A unique typed copy is in Eleutherian Mills Historical Library, Wilmington, Delaware, cataloged

as Curtis Publishing Co., *Automobiles: Gasoline Pleasure Cars* (Philadelphia, 1914). See p. 453 for the attitudes of bankers.

62. Harry Braverman, *Labor and Monopoly Capital: The Degradation of Work in the Twentieth Century* (New York, 1974), chap. 5, "The Primary Effects of Scientific Management." The Lynds, in *Middletown in Transition* (note 34), pp. 64–65, note the deterioration of skills in manufacturing industries.

63. Musson, "Joseph Whitworth and the Growth of Mass-Production Engineering" (note 51). There is no question that the American system drew heavily at many points upon British technical knowledge and machines. Yet the same technical information was used differently in Britain and America. Thus American technology became something that British technology was not becoming. The Portsmouth machinery for making ships' pulley blocks, promoted by Bentham, designed by Brunel, and built before 1808, consisted of a series of 43 special-purpose machines used to produce 130,000 blocks a year at an estimated 90 percent saving of labor. Here is the American system in full cry, but the scheme was not adapted in England to other manufactures. Charles Singer et al., eds., *A History of Technology*, 5 vols. (New York, 1954–58), 4: 426–28. Otto Mayr, in the exhibition "A Nation of Nations" at the National Museum of American History, makes a plausible case for the international origins of elements of the American system. The sewing machines, typewriters, and automobiles originated in Europe; in America they were transformed as they were adapted to large-scale manufactures. See *A Nation of Nations*, ed. Peter C. Marzio (New York, 1976), pp. 508–63.

64. Davies, *Peacefully Working to Conquer the World* (note 53), p. 79.

BRITISH ORIGINS

A. E. Musson

This essay is concerned with demonstrating that mass-production engineering—the production of standard interchangeable parts by means of power-driven machine tools—did not, as is generally believed, originate in the United States but in Europe and especially in Great Britain, in the late eighteenth and early nineteenth centuries, and that American advances in this field occurred mainly from the midnineteenth century onward.

The Industrial Revolution in Britain created a rapidly growing demand for prime movers and machinery, quantity production of which could be achieved only by development of machine tools and precision engineering. Though there were developments in other European countries, as by Polhem in Sweden and Blanc in France for mass production of firearms, the major advances were made in Britain by engineers such as Bramah, Maudslay, Clement, and others in London, and by Roberts, Fairbairn, Nasmyth, Whitworth, and others in Manchester. These engineers developed and eventually mass-produced self-acting lathes, drilling and boring engines, planers, shapers, and other machine tools which in the first half of the nineteenth century made possible standardized mass production of textile and other machinery, using measuring machines, gauges, and templates, and leading to something resembling assembly-line production.

During this period the United States generally lagged well behind, drawing know-how, machinery, and skilled immigrants from Britain. Though mass production was developed in firearms and woodworking manufactures, by Whitney, North, Hall, and others, it was not until the 1840s that the American engineering industry began to grow rapidly. From then on the United States started to assume leadership not only in firearms manufacture, but also in production of clocks and watches, locks, sewing machines, typewriters, and agricultural machinery, and in the associated development of new machine tools such as millers and turret lathes. Yet the reports of British observers in the 1850s show that the so-called "American system of manufactures" had developed only in a few branches of light engineering, while British engineers and machine tools generally remained superior. The conventional view—that mass-production methods did not exist in Britain until introduced from the United States for firearms manufacture in the 1850s—is largely erroneous. In the first half of the nineteenth century, before America's massive industrial expansion, the traffic across the Atlantic was predominantly the other way, though "Yankee ingenuity" was already becoming apparent and was soon to establish American engineering leadership.

One would naturally expect to find that machine tools, machinemaking, and mass production were developed originally in Great Britain, the country that first experienced an industrial revolution and became "the workshop of the world." Indeed that very term originated in connection with the rapid growth of the British machinemaking industry and the British campaign for free trade and expansion of export markets in the second quarter of the nineteenth century.[1] America, on the other hand, was relatively backward during this period, drawing much of its technology from Great Britain. It would therefore seem improbable that by the 1850s a unique "American system of manufactures," on standardized mass-production lines, could have developed to the extent portrayed in the traditional historical interpretation of the reports of contemporary British observers who used that term. Yet according to this now-widespread interpretation, the Americans pioneered in developing precision manufacture of standard interchangeable machine parts by means of machine tools and gauges, and these methods were then gradually introduced into Britain and Europe from the 1850s onwards. In the United States, we are told, economic opportunities, relative labor costs, markets, and social attitudes were such as to encourage a more enterprising use of labor-saving machinery than in Britain, with its more specialized consumer demands, relatively cheap labor, conservative craft practices, and trade-union restrictions. The common acceptance of this view is evident in the original prospectus of this symposium, which referred to "the American system of manufactures" as "the technology of mass production... pioneered in the United States in the first half of the nineteenth century."

Britain's relative backwardness in development of this mass-production technology has been repeatedly asserted. Roe, despite his knowledge of British machine-tool makers, ascribed the development of "interchangeable manufacture" to the United States.[2] Rothbarth declared that nineteenth-century British industry "never began to become efficient in mass-production methods."[3] Sawyer emphasized the social factors underlying American superiority in this field.[4] Habakkuk's comparative study of American and British technology also was based on the assumption of

American leadership in labor-saving mechanization in the first half of the nineteenth century, despite his awareness of early British pioneering.[5] Rosenberg, while acknowledging America's debt in iron and steam technology, has likewise reiterated how and why mass production originated in the United States rather than in Britain.[6]

This view, however commonly accepted, is open to serious question. My own researches over the past twenty years or more have demonstrated the early development of machine tools and mass-production engineering in Britain from the late eighteenth century onward,[7] and Saul also has shown how "seriously misleading" is the exclusive emphasis on American pioneering and the assumption "that in this area of engineering technology British industry lagged particularly badly."[8] Among American scholars, too, there has been some critical reappraisal. Woodbury, for example, has not only demonstrated the European origins of machine tools and standardized manufacture, but has also demolished some of the legends relating to Eli Whitney.[9] Temin, while tending to accept the view of midnineteenth-century American technological superiority, has emphasized the very narrow range of industries—mostly woodworking, firearms, and hardware—on which British visitors to the United States in the 1850s reported, and has stressed "the lack of evidence about whole areas of American industry and the known American backwardness in the production and fabrication of iron."[10] David has cast similar doubts on Habakkuk's labor-scarcity thesis and on the travelers' reports themselves, stressing the lack of reliable information about comparative technological progress in Britain and America.[11]

It certainly seems that the early pioneering achievements of Whitney, Simeon North, and others have been exaggerated. In fact most of the evidence of American leadership dates from the midnineteenth century onward. It was in the third quarter of the century that Burn—rightly, in my view—traced "the genesis of American engineering competition."[12] Habakkuk, while emphasizing that American superiority had become marked by midcentury, recognizes that "the idea of interchangeable parts was not American but French in origin";[13] that Brunel and Bentham in England hit upon the idea of interchangeability at about the same time as Whitney and North in America; and that "most of the general machine tools had been invented by British engineers between 1775 and 1850—boring machines, engine lathes, planers, shapers, steam hammers, and standard taps and dies."[14] Habakkuk goes on, of course, to point out that "the most important new machine tools, particularly the milling machines and turret lathes, were developed in America," but these were not brought into general use until the *second half* of the century, and milling cutters had been developed in England as well as in America.

Habakkuk admits that "American excellence in the production of machine tools did not become evident until after the 1850s and later," and that while "even in the 1850s the Americans had, for a number of operations, machine tools more specialized than those available in England," these were as yet mostly confined to "woodworking and small arms." In the preceding period, indeed, the Americans were actually importing British machine tools, supplied by Whitworth and other makers.[15]

Rosenberg has been more forthright than Habakkuk in declaring that, although America borrowed iron and steam technology from Britain, the system of standardized mass production was pioneered and developed in the United States. This "American system of manufacturing," he asserts, was unknown in Britain until the Great Exhibition of 1851, when it was first demonstrated to the British public in American firearms manufacture and was thereafter imported into Britain, notably in the Enfield Ordnance Factory, established in 1854–55. This "marked the beginning of the movement of mass-production techniques from the United States to Europe."[16] The great weight of Rosenberg's evidence, however, like Habakkuk's, relates to the second half of the century; indeed, he dates the really significant growth of the American machine-tool industry and standardized manufacture only from about 1840.[17] Prior to that, he points out, machinemaking techniques and machine tools were only beginning to evolve as adjuncts to textile factories such as the Amoskeag Manufacturing Co., in Manchester, New Hampshire, and the Lowell Mills in Lowell, Massachusetts, which started manufacture and sale not only of textile machinery but also of waterwheels, steam engines, mill machinery, machine tools (lathes, planers, boring machines), and eventually, from the 1830s onward, railway locomotives. At the same time, more specialized machine production was also evolving in firearms manufacture and woodworking. Nevertheless, the widespread development of mass production, with milling machines, turret lathes, and precision grinders, did not occur until the second half of the century, with the growing manufactures of clocks and watches, locks, sewing machines, typewriters, agricultural machinery, and, later, bicycles and automobiles. In Britain by contrast, according to Rosenberg, there was a long lag in the development of laborsaving mechanized mass production, partly because of England's alleged disadvantages as an "early starter"—skilled handicraft methods, "a wide variety of standards and specifications," "technological interrelatedness," and costs of scrapping and rebuilding—and partly on account of more diverse consumer tastes and markets, and stronger, more conservative trade unions.

With Rosenberg's thesis and evidence for the second half of the century, with his emphasis on the fundamental importance of the machine-tool industry for diffusion of "machinofacture" to an ever-widening range of industries, and with his emphasis on growing American leadership in that period, I entirely agree. But I cannot agree with him about the earlier period. If, as he recognizes, America was borrowing so heavily from Britain in coal, iron, and steam technology in the first half of the nineteenth century; if America remained so long reliant on wood and water for the construction material, fuel, and power for its industries; if America was so relatively backward in "modern" technology in that period; is it not then probable that America also originally borrowed machine tools and machinemaking from the "workshop of the world"?

As Woodbury has shown, the idea of standardized mass production of firearms originated in Europe, with the early-eighteenth-century pioneering of Christopher Polhem in Sweden, and that of the Frenchman, Honoré Blanc, later in the century. Undoubtedly, with the growing market for firearms in the United States, this "interchangeable manufacture" using precision machine tools such as milling machines was vigorously developed in the late eighteenth and early nineteenth centuries, not only by Whitney, but also by Simeon North, John Hall, the Springfield Armory and Harpers Ferry Armory, and later by Robbins and Lawrence, the Ames Manufacturing Co., and Colt's armory. Nevertheless, a great deal of skilled labor and hand filing and fitting were still required and, as Saul has emphasized, American mass production of firearms progressed slowly and not altogether successfully in the first half of the nineteenth century; indeed, "interchangeable manufacture was to all intents and purposes abandoned" at Springfield and at the Colt factory in the late 1840s.[18] There is no doubt, however, that in this particular field, and also in woodworking machinery, the United States had attained world leadership by the midnineteenth century, partly because of the market forces previously mentioned and partly because of relative factor costs—particularly scarcity of skilled labor and abundance of cheap wood. Rosenberg is therefore quite right in contrasting "the American system of manufactures" in the firearms industry with the traditional handicraft methods that long predominated in Birmingham; in this industry, undoubtedly, the Enfield and Colt factories were remarkable innovations.

The Birmingham small-arms trade, however, may also be contrasted to other British industries almost totally overlooked by Rosenberg, in which precision engineering and mass production had been vigorously developing since the late eighteenth century. These were particularly important in the making of steam engines, textile machinery, railway locomotives, and above all machine tools, which were as fundamental to British as to American "machinofacture." It is evident, indeed, that the textile-machine shops of New England, which Rosenberg refers to as crucial in the early development of heavy general-purpose machine tools, actually acquired their tools and techniques from old England, whence their powerdriven textile machinery had originated.[19] The Lowell machine shop, for example, not only copied and adapted Lancashire cotton-spinning machinery, but also imported English machine tools and files; except for powered lathes, its machinemaking methods remained relatively primitive, with much blacksmith work and filing, until about 1840, when "great improvements in metal-working skills were brought about by the importation from England of new machine tools," such as Whitworth planing, drilling, and boring machines.[20] Similarly, from the early 1830s onward, the Lowell shop copied English locomotive-building practices.[21]

During the first half of the nineteenth century, cotton manufacture was America's greatest industry, and G.S. Gibb emphasizes that for much of that period "the manufacture of textile machinery appears to have been America's greatest heavy goods industry," first in size and value of product among all the metalworking industries, and playing a "crucial" role in the diffusion of engineering skills: "From the textile mills and textile-machine shops came the men who supplied most of the tools for the American Industrial Revolution. From these mills and shops sprung directly the machine tool and locomotive industries, together with a host of less basic metal-fabricating skills." [22] Yet in this evolution of the American textile-machine making, machine-tool, and other metalworking industries, Gibb points out, the United States had remained "vitally dependent on England for models of new machine tools and textile machinery." He continues:

> Throughout the entire period under consideration [1813–53] and particularly in the 1840s and 1850s American machine-shop owners were sending their ranking mechanics to England to purchase, to observe, and to learn. . . . In 1853 English metal-skills still surpassed those to be found in America . . . English machine-tools far surpassed their American counterparts in design and workmanship. English textile machines exhibited a similar superiority.

Only from the 1840s onwards did specialized machine-tool makers emerge in the New England industry.[23]

Gibb's findings confirm those of Victor S. Clark's massive history of American manufactures. During the first half of the nineteenth century, Clark emphasized, "British industrial technique was . . . constantly so far ahead of that of the United States that our [American] manufacturers always were in the position of learners from the old country." [24] Only "after 1840" was there very significant "improvement in metal-working tools and machinery," including "the use of interchangeable mechanism." [25] During this half-century, the United States "remained in engineering tutelage to the older country," as demonstrated by the general inferiority of American heavy machine tools to British in the early 1850s.[26] Only in the light-metal manufactures such as firearms and nails, and also in woodworking, had American superiority been established, with the development of specialized machine tools for use in highly subdivided manufacturing processes.[27] In textile-machine making, locomotive manufacture, and general heavy engineering, Britain remained ahead.

In these British industries precision engineering—with "self-acting" machine tools, gauges, templates, and standardization—had long been developing. Joseph Whitworth has often been regarded as the great British pioneer in precision engineering and in his advocacy of "the American system of manufactures" in the 1850s. But, as the present writer has shown, this view of Whitworth (propagated mainly by himself) is largely mistaken: while undoubtedly a great engineer, his most outstanding contributions were in improving, perfecting, and publicizing techniques that had been developing even before he was born.[28] Though much of the machinery of the early Industrial Revolution was made of wood, leather, and castings of iron and brass, machine tools had become increasingly important in the late eighteenth and early nineteenth centuries for cutting and shaping metal parts.[29] The origins of the lathe and slide-rest are to be sought centuries earlier, but Maudslay had developed an all-iron "self-acting" lathe, combining slide-rest with change-gears and power-driven lead-screw, for heavy industrial use, by the end of the eighteenth century.

Henry Maudslay (1771–1831). (From an original drawing in the collections of the Division of Mechanical and Civil Engineering, National Museum of American History, based on a contemporary portrait; Smithsonian Institution Neg. No. 81–189.)

Henry Maudslay's treadle-powered bench lathe, ca. 1800–1810. (Crown copyright, Science Museum, London.)

Much earlier in that century, clockmakers such as Hindley and Wyke had been making wheel- or gear-cutting engines, as well as lathes and other tools, which they had supplied to Smeaton, Watt, Boulton, Wedgwood, and others; such clockmakers and instrumentmakers played a key role in the development of precision engineering, especially with what was referred to as "their more accurate and scientific mechanism," which could be used in the "clockwork" of textile and other machinery.[30]

There are numerous references to lathes, drills, and boring and wheel-cutting engines, together with "unerring gauges," in the growth of Lancashire engineering in the last quarter of the eighteenth century, when scores of firms sprang up to make spinning machinery, millwork, waterwheels, and steam engines.[31] In the manufacture of steam engines, as is well known, there was an increasing need for accuracy, which led to the development of improved boring engines by Smeaton and Wilkinson. Wilkinson produced in 1774 what Roe described as "probably the first metal-working tool capable of doing large heavy work with anything like present-day accuracy."[32] Valves and other precision parts also were made with the aid of machine tools. In the early nineteenth century, while Whitworth was still in his boyhood, the earliest metal-planing machines were being invented or developed by several engineers, including Fox of Derby, Murray of Leeds, Roberts of Manchester, Clement and Rennie of London, and Spring of Aberdeen.

These early machine tools were soon followed by others, as toolmaking increasingly tended to become a specialized trade, distinct from textile-machine making, steam-engine making, and other branches of mechanical engineering. Wyke and other toolmakers of the second half of the eighteenth century were followed in the early nineteenth by more famous engineers specializing in the manufacture and sale of machine tools, including Maudslay, Clement, and Holtzapffel in London, Fox of Derby, Murray of Leeds, and Roberts of Manchester, followed later by Nasmyth, Whitworth, and others, together with similar machine-tool makers in other centers such as Glasgow. Not only were lathes and boring, drilling, gear-cutting, and other early machine tools greatly improved, but new ones were invented for such operations as planing, slotting, shaping, milling, punching, and shearing. British mechanical engineering was, in fact, transformed in the second quarter of the nineteenth century by the development of these power-operated machine tools, as evidenced by the report of the Select Committee on Exportation of Machinery in 1841,

Model of John Wilkinson's waterpowered boring mill of 1774, the first metalworking tool capable of precision in large-scale work. (Crown copyright, Science Museum, London.)

which declared that such tools had "introduced a revolution in machine-making, and tool-making has become a distinct branch of mechanics, and a very important trade, although twenty years ago it was scarcely known."[33] The committee pointed out that power-driven "self-acting" machine tools, costing £100 to £2,000 each and operated by semiskilled machine-minders, could achieve a much greater output of more accurately manufactured, standardized machinery at far lower cost than a generation previously.[34] These findings, based on the evidence of reputable engineers, were confirmed by the recollections of others such as Fairbairn and Nasmyth. Both emphasized the comparatively rudimentary state of engineering about 1815, when there were no planing, slotting, or shaping machines, and, except for some primitive lathes and drills, machinery was largely made by hand. By about 1840, however, as they and other observers such as Andrew Ure reported enthusiastically, "self-acting" machine tools had brought about a "revolution" in machinemaking, almost entirely displacing handicraft skills and producing precisely engineered, interchangeable parts.[35]

From the late 1830s onwards, these developments, with increasing emphasis on standardization, were certainly extended by Whitworth, but they had long been under way: early examples of standardized mass production, using specialized machine tools, include the Wyatt brothers' factory in Staffordshire for manufacturing wood-screws, and the Taylors' plant for making ships' blocks at Southampton, both dating from the 1760s.[36] Much more significant were the achievements of Bentham, Brunel, and Maudslay at the turn of the nineteenth century, with their famous blockmaking machinery for the Portsmouth naval dockyard,[5] following the equally important but less well-known achievements of Bramah and Maudslay in the manufacture of locks.[37] In both cases a series of special machine tools was devised for making the various parts, but, whereas the former was woodworking machinery, the latter consisted of metal-cutting machines to mass-produce locks "with the utmost precision," as John Farey, the London engineer, later described.[38] With their slide-lathes and milling cutters, as Roe acknowledged, Bramah and Maudslay "laid the foundations for the metal-cutting tools of today."[39]

Roe went on, however, to state that this "interchangeable system of manufacture" developed by Bramah, Brunel, and Maudslay had "little or no influence on the general manufacturing of the country," so that there was complete stagnation in this field until 1855, "when England... imported from America the Enfield gun machinery and adopted what they [the English] themselves styled the 'American' interchangeable system of gunmaking."[40] Roe's assertion has been repeated uncritically by Habakkuk, who states that because of the abundance of cheap labor in En-

*Brunel and Maud-
slay's block-shaping
engine (1804), one of
the machines built for
the Portsmouth Naval
Yard. (Crown copy-
right, Science Muse-
um, London.)*

gland, "the idea of interchangeable parts" developed by Bramah, Brunel, and Maudslay "was not taken up by any other English industry."[41] Ames and Rosenberg have been equally uncritical in accepting these statements by Roe and Habakkuk.

It is clear, however, that this view of stagnation in British mechanical engineering during the first half of the nineteenth century is grossly mistaken. It would, indeed, be extraordinarily surprising if the labor-saving incentives Habakkuk recognizes in Bentham, Brunel, and Maudslay had been so completely lacking in other British engineers during the next half-century. In fact, as we have just seen, there were remarkable laborsaving developments in precision engineering, machine tools, and standardized machinemaking during that period. Maudslay had developed not only machine tools, but also accurate measuring machines (used as bench micrometers), standard gauges, true planes, and standard screw-threads, thereby laying the foundations for standardized production, which were improved upon both in London and in the provinces by other famous engineers, many of whom had been trained in his workshop. John Clement, for example, after being chief draughtsman first for Bramah and then for Maudslay, set up on his own in 1817, further developed the lathe and planing machine, and continued Maudslay's work in manufacturing taps and dies based on his standard screw-threads. He also devised special machine tools for making the numerous standard parts in Babbage's famous calculating machine in the early 1830s.[42] About the same time, Bryan Donkin, another notable London engineer, was similarly engaged on the Fourdrinier papermaking machinery. As the young American engineer George Sellers observed in 1832: "This was the first instance I had seen where making the component parts of machinery interchangeable had been reduced to an absolute system, that is now so universally practised by all first-class machinists."[43]

Maudslay's ideas were spread in the provinces by such later-famous engineers as Roberts, Fairbairn, Nasmyth, Whitworth, and Muir—all of whom had acquired experience in his workshop before establishing themselves in Manchester. With the prodigious growth of the cotton industry, that city was to eclipse London as the great center of British engineering in development of standardized mass production. Peel, Williams & Co., for example, probably the biggest engineering firm in Manchester during the first two or three decades of the nineteenth century—producing textile machinery, boilers, steam engines, hydraulic presses, and eventually railway locomotives, as well as their own machine tools—had developed by 1806, if not earlier, "a large and extensive series of [gear] wheel models." These they advertised widely in printed catalogs, listing the various sizes of gear-wheels, with their diameter, number of teeth, pitch, and breadth of cog, together with models of other engineering products. Here is clear evidence of standardized mass production of engineering goods in anticipation of demand.[44]

Another, more famous Manchester engineer, Richard Roberts, became preeminent in the manufacture of machine tools.[45] In addition to his planing machine, he produced improved lathes and gear-cutting, drilling, and slotting machines, which, together with gear-wheels and screws, he was advertising for sale by the early 1820s. He was able to develop mass-production techniques, using gauges and templates, first in the manufacture of textile machinery (including his famous "self-actor" mule, patented in 1825) and later in locomotive building. Andrew Ure was greatly impressed by Roberts's manufacture of standardized textile machinery in 1835:

Where many counterparts or similar pieces enter into the spinning apparatus, they are all made so perfectly identical in form and size, by

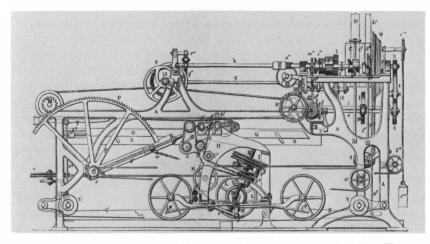

"Self-actor" mule of Sharp and Roberts, patented in 1825. (From Andrew Ure, The Cotton Manufacture of Great Britain, *vol. 1; Smithsonian Institution Neg. No. 81–196.)*

the self-acting tools, such as the planing and key-groove cutting machines, that any one of them will at once fit into the position of any of its fellows in the general frame.

For these and other admirable automatic instruments, which have so greatly facilitated the construction and repair of factory machines, and which are to be found at present in all our considerable cotton mills, this country is under the greatest obligations to Messrs. Sharp, Roberts, and Co. of Manchester.[46]

Though manufacturing to gauge had developed in Lancashire machine-making many years earlier, as we have previously noted, there is no doubt that Roberts (like Whitworth after him) did much to spread such standards. In locomotive manufacture from the 1830s onwards, he similarly applied self-acting machine tools, together with what Samuel Smiles called his "system of templets and gauges, by means of which every part of an engine or tender corresponded with that of every other engine or tender of the same class." This, Smiles declared, "was as great an improvement as Maudslay's system of uniformity in other descriptions of machinery."[47]

John G. Bodmer was an equally outstanding engineer, a Swiss immigrant who set up machine shops in Bolton in the mid-1820s and in Manchester in the 1830s.[48] Highly versatile in the development of machine tools, he had produced a measuring machine and "a complete system of distinctly marked and accurately executed internal and external gauges of various kinds" for making standardized textile machinery.[49] Indeed, during the later Napoleonic wars, long before Bodmer came to England, he had developed the standardized manufacture of small arms, for which he "invented and successfully applied a series of specialised machines by which the various parts—more especially the lock—were shaped and prepared for immediate use, so as to ensure perfect uniformity and to economise labour."[50] Later, in England, contemporaneously with Roberts, he developed machine tools and applied templates and gauges to the manufacture of textile and other machinery.

James Nasmyth further developed these techniques when he established his famous Bridgewater Foundry at Patricroft, Manchester, in 1836, deciding "to have such as planing machines and lathes, etc. etc., all ready to supply."[51] These "ready made" machine tools would be manufactured by something resembling modern assembly-line production, the whole works being laid out "all in a line," or on "what may be called the straight line system," so that "the work, as it passes from one end of the foundry to the other, receives in succession, each operation which ought to follow the preceding one."[52] Nasmyth thereby produced stocks of standardized machine tools of various sizes which he advertised for sale in printed catalogs, while at the same time bringing out new or improved tools such as grooving, shaping, and milling machines, as well as lathes, planers, drills, boring machines, gear-cutting and screw-cutting machines, and, of course, his famous steam hammers. All these machines were "self-acting." Their standardized mass production—for use in making standardized machinery for the mass production of standardized "end products"—obviously heralded the modern machine age.[53]

Nasmyth himself mass-produced equipment such as steam-pumping engines and hydraulic presses, and in a description of the Bridgewater Foundry in 1856 we read of the assembly, ready for sale, of "a regiment of donkey pumps all in a line.... marshalled columns of ambidextrous lathes and grooving machines, hydraulic presses ... [and] wrought iron cranes," together with steam hammers and other products.[54] Nasmyth also produced standardized batches of locomotives on the same system, for various railway companies, from the late 1830s onward.

Many other engineers in Manchester and elsewhere in Lancashire were similarly stimulated by the revolution in the cotton industry and by the new railway age. In Manchester and Salford alone, according to Slater's directory of 1845, there were almost a hundred listings for millwrights, engineers and machinists, machinemakers, and ironfounders. Classified lists in the trade directories during the first half of the nineteenth century evidence a growing number of specialized toolmakers and machinemakers. Testimony taken by the Select Committee of 1841 also indicates

James Nasmyth (1808–1890) in carica-
ture. (From Punch, *vol. 84 [1883]; Smith-*
sonian Institution Neg. No. 81–188.)

that the manufacture of machine tools was becoming a specialized branch of engineering, distinct from the manufacture of steam engines and mill-work on the one hand and textile-machine making on the other, though many firms remained general engineers.

Similar though less striking developments were occurring in other industrial areas of Britain during this period. In the manufacture of locomotives, as in the manufacture of machine tools and textile machinery, as Saul has pointed out, "standardization and interchangeability were certainly not confined to, nor originated by, American builders." Following the example of Roberts and Nasmyth, Daniel Gooch insured interchangeability for his first series of 142 standard locomotives for the Great Western in the early 1840s by sending out precise specifications and templates to the various firms employed in building them.

It is abundantly clear that Whitworth was by no means the great pioneer of standardized mass production in Britain that he proclaimed himself to be—and as which he has been accepted by historians. Nevertheless, he did a great deal to publicize and spread these engineering principles, by his own manufactures—particularly the machine tools he displayed at the 1851 and 1862 London exhibitions, together with his measuring machines, true planes, and standard gauges and screws; by his papers to the British Association, the institutions of Civil and Mechanical Engineers, and other bodies from the late 1830s onward; and by his evidence to parliamentary committees in the 1850s on such subjects as American manufactures, small-arms production, and standard measures.[56] A strikingly successful demonstration of his methods was the construction of engines for ninety steampowered gunboats required by the Admiralty for the Crimean War in 1855: these were completed in ninety days, "a feat which had been rendered possible only by the general adoption of the Whitworth standards, and by the consequent power of obtaining from different engineering shops the separate parts of the engines," manufactured "with such accuracy that the several parts could be put together at once for delivery, without having to undergo any further fitting for completing the erection of the engines."[57]

Whitworth was able to claim—and his claims were supported by contemporary authorities in *The Engineer* and *The Times*—that his system of standardization had been widely adopted by the late 1850s, both in private and in government engineering establishments.[58] But, as we have seen, this success was based on the work of many forerunners from the late eighteenth century onward, and so it is clearly erroneous to maintain that precision engineering and standardization were virtually unknown in Britain before the introduction of "the American system of manufactures" in the 1850s. I would suggest instead that research should be directed toward discovering to what extent American developments during the first half of the nineteenth century were based upon the transfer and diffusion of British machine-tool and machinemaking techniques, and to what extent they were of American origin.

It is also clear that the Habakkuk thesis—that there was a great lag in British engineering development because of abundant supplies of relatively cheap labor—is largely untenable. Indeed, Habakkuk's own arguments on this issue are inconsistent, since he recognizes that compared to the United States skilled engineering labor in Britain was relatively scarce and dear.[59] Rosenberg has subsequently added emphasis to this point by showing that wages of the most skilled British machinemakers were not only relatively but even absolutely higher than those of comparable American workers.[60] It is clear that from the late eighteenth century onward, British engineering employers had experienced a considerable scarcity of skilled workers, and that high wages and trade-union apprenticeship and machine-manning restrictions were a constant source of complaint.[61] Consequently, contrary to Habakkuk's thesis, employers had constantly sought laborsaving innovations, especially by introducing "self-acting" machine tools which could be tended by unskilled or semiskilled laborers at lower wages. And not only were they substituting unskilled for skilled labor, they also were substituting capital (machinery) for labor, since more such machines could be managed by a single operative.

These developments were associated with increasing subdivision and specialization of labor, as the general craft skills of the millwright gave way to those of patternmakers, moulders, turners, borers, and fitters. Machine-minders could be employed in repetitive production of standardized parts on specialized machine tools, without any apprenticed training. These increasingly specialized operations in larger factories necessitated planned production, careful coordination, and disciplined control. Thus in Fairbairn's Manchester engineering works in the late 1830s a contemporary described how the "550 to 600 hands [were] employed in the various departments," affording a striking "example of practical science." "In every direction of the works the utmost *system* prevails, and each mechanic appears to have his peculiar description of work assigned with the utmost economical subdivision of labour." The various specialized groups of workers "appear to attend to their respective employments with as much regularity as the working of the machinery they assist to construct." [62] Mechanized mass production obviously involved profound organizational as well as technological changes.

These developments, however, were not brought about without a good deal of human friction. The employment of semiskilled, specialized machine-minders in disciplined factory production roused strong trade-union opposition, culminating in the famous 1852 strike. But such opposition could not prevent (though it may have delayed) technological change and gradual labor dilution; indeed, as Habakkuk recognizes, there is much evidence that it was a direct cause of laborsaving mechanization. [63]

In his famous report on American manufactures, Whitworth emphasized that American workers were much more willing than British workers to cooperate in technological innovation. Because they were better educated and because labor was less abundant, there was less fear of labor redundancy, more appreciation of the benefits of mechanization, an absence of apprenticeship restrictions, more social mobility, and less class hostility. In many manufactures, other than machinemaking, it would indeed appear that labor was more abundant and cheaper in England than in the United States, where standardized mass production therefore developed more rapidly in the second half of the nineteenth century. Sociocultural factors of the kind referred to by Whitworth, as well as relative factor costs and the growth of a more homogeneous mass market, no doubt contributed to this American manufacturing development, especially in branches of light engineering such as those in which Burn,

Rosenberg, and others have described the growth of American mass production during the third quarter of the nineteenth century. These factors largely explain why and how the United States caught up with and surpassed Great Britain in such manufactures after 1850, though, as Saul and Floud have pointed out, American supremacy developed neither so suddenly nor so completely as previously portrayed; indeed, in several branches of heavy engineering, as well as in some light engineering such as bicycle manufacture, Britain maintained supremacy up to 1914.[64]

Nevertheless, the observations of Whitworth and others in the 1850s have generally been interpreted as supporting American engineering superiority. The fact is, however, that their observations contain clear indications that, except in woodworking, small arms, and a few other light metal manufactures, British engineering was definitely more advanced.[65] Whitworth had been struck most by the Americans' superiority in woodworking machinery, whereas he considered them "not equal to us in the working of iron."[66] He clearly affirmed Britain's leadership in the making and use of heavy machine tools, declaring that even though a higher proportion of American lathes had self-acting slide-rests, the "engine tools employed in the different works are generally similar to those which were used in England some years ago, being much lighter, and less accurate in their construction, than those now in use, and turning out less work in consequence." He saw planing and drilling machines in common use, "but there are comparatively very few horizontal or vertical shaping machines, and a considerable amount of hand labour is, therefore, expended on work which could be performed by machines much more economically."[67] Ordnance officials who visited the United States the following year also found that, "as regards the class of machinery usually employed by engineers and machine makers, they are upon the whole behind those of England."[68]

On the other hand, there is no doubt that in the mechanization of various light manufacturing processes the Americans were by this time taking the lead, especially in woodworking, firearms, locks, clocks, and watches, and various hardwares such as nails, and were about to develop a similar superiority in sewing machines, typewriters, agricultural machinery, and, most significantly, in new machine tools such as milling and grinding machines and turret lathes employed in manufacturing these products. It was this application of highly specialized machine tools to a sequence of manufacturing operations, most notably in woodworking and the making of small arms, that so strongly impressed British observers in the United States in the early 1850s as a distinctively "American system of manufactures" different from the British use of general machine tools. American progress in the mass production of firearms similarly struck

James Nasmyth on visiting Colt's Pimlico factory at this time, contrasting with the "traditional notions, and attachments to old systems" that prevailed in comparable British manufactures.[69]

In such developments, as Whitworth remarked, there is no doubt that after the pioneering of Bramah and Maudslay, British engineering had lagged. The basic principles of such a manufacturing system were well known to British engineers, as Nasmyth pointed out. Like Whitworth, he attributed their relatively slow application mainly to craft conservatism and trade-union restrictions, such as apprenticeship and manning regulations. No doubt these restrictions were inhibiting (they contributed, for example, to Nasmyth's early retirement),[70] but there were other economic factors that help to account for Britain's slower adoption of mechanized mass production in such manufactures—factors of the kind emphasized by Habakkuk and Rosenberg relating to markets, factor costs, and resource endowments: the waste of relatively dear timber by woodworking machines; a smaller, more varied and fluctuating market for firearms; together with relatively abundant and cheap skilled labor for making these and other small metal manufactures such as nails and clocks. From an economic point of view, therefore, British firms in these industries were not entirely irrational in their more prolonged adherence to handicraft methods. At the same time, as we have seen, in general heavy engineering and machinemaking, Britain had pioneered the development of machine tools, precision techniques, and standardization, which the United States had borrowed and adapted. It was only from the mid-nineteenth century that America began to repay her earlier debts.

Notes

1. A. E. Musson, "The 'Manchester School' and Exportation of Machinery," *Business History* 14 (1972): 17–50.

2. J. W. Roe, "Interchangeable Manufacture," *Newcomen Society Transactions* 17 (1936–37): 165–74; *English and American Tool Builders* (New Haven, Conn., 1916).

3. E. Rothbarth, "Causes of Superior Efficiency of U.S.A. Industry as Compared to British Industry," *Economic Journal* 54 (1946): 383–90.

4. J. Sawyer, "The Social Basis of the American System of Manufacturing," *Journal of Economic History* 14 (1954): 361–79.

5. H. J. Habakkuk, *American and British Technology in the Nineteenth Century* (Cambridge, 1962).

6. Several of Rosenberg's articles relating to this subject are included in his *Perspectives on Technology* (Cambridge, 1976). See also his *Technology and American Economic Growth* (New York, 1972); *The American System of Manufactures* (Edinburgh, 1969); and, as coauthor with E. Ames, "The Enfield Arsenal in Theory and History," *Economic Journal* 78 (1968): 827–42.

7. Several of my articles, together with others written in collaboration with Eric Robinson, have been included, in revised form, in *Science and Technology in the Industrial Revolution*, ed. A. E. Musson and E. Robinson (Manchester, 1969). See also my introduction to the reprint of *The Life of Sir William Fairbairn*, ed. W. Pole (Newton Abbot, 1970); and "Joseph Whitworth and the Growth of Mass-Production Engineering," *Business History* 17 (1975): 109–49.

8. S. B. Saul, ed., *Technological Change: The United States and Britain in the Nineteenth Century* (London, 1970), chap. 5.

9. R. S. Woodbury, *Studies in the History of Machine Tools* (Cambridge, Mass., 1972); "The Legend of Eli Whitney and Interchangeable Parts," *Technology and Culture* 1 (1959–60): 235–53.

10. P. Temin, "Labor Scarcity and the Problem of American Industrial Efficiency in the 1850s," *Journal of Economic History* 26 (1966): 277–98.

11. P. A. David, *Technical Choice, Innovation and Economic Growth: Essays on American and British Experiences in the Nineteenth Century* (Cambridge, 1975), chap. 1.

12. D. L. Burn, "The Genesis of American Engineering Competition, 1850–1870," *Economic History* 6 (1931): 292–311.

13. It can be traced back, in fact, to the Swedish engineer, Christopher Polhem, in the early eighteenth century, as well as to the French experiments with firearms later in that century. A. P. Usher, *A History of Mechanical Inventions* (rev. ed., Cambridge, 1954), pp. 376–79; Roe, "Interchangeable Manufacture" (note 2).

14. Habakkuk, *American and British Technology* (note 5), pp. 104–5, 116. The first half of the nineteenth century was, in fact, "the great age of English machine-tool invention, the age of Bramah, Maudslay, and Nasmyth" (ibid., p. 117).

15. Ibid., pp. 105–6. On early British milling machines, see Musson, "Joseph Whitworth" (note 7), pp. 137–38.

16. Rosenberg, *Technology and American Economic Growth* (note 6), chap. 4; *The American System of Manufactures* (note 6), intro.; and Ames and Rosenberg, "The Enfield Arsenal in Theory and History" (note 6).

17. Rosenberg, "Technological Change in the Machine Tool Industry, 1840–1910," *Journal of Ecnomic History* 23 (1963): 414–46 (reprinted in *Perspectives on Technology* chap. 1).

18. Saul, *Technological Change* (note 8), pp. 7–8, based on C. H. Fitch, "Report on the Manufactures of Interchangeable Mechanism," in *Report on Manufactures of the United States at the Tenth Census* (Washington, D. C., 1883),

and F. J. Deyrup, *Arms Makers of the Connecticut Valley* (Northampton, Mass., 1948).

19. G. S. Gibb, *The Saco-Lowell Shops: Textile Machinery Building in New England, 1813–1949* (Cambridge, Mass., 1950); T. R. Navin, *The Whitin Machine Works Since 1831* (Cambridge, Mass., 1950). The heavy American dependence on British textile technology has been comprehensively demonstrated by David J. Jeremy, "The Transmission of Cotton and Woollen Manufacturing Technologies Between Britain and the U.S.A. from 1790 to the 1830s" (Ph.D. diss., London School of Economics and Political Science, University of London, 1978). Though Jeremy shows that there were some reverse flows, he states (p. 292) that "before 1840 relatively few American innovations were adopted in England."

20. Gibb, *The Saco-Lowell Shops* (note 19), pp. 81–82.

21. Ibid., pp. 92–96.

22. Ibid., p. 179.

23. Ibid., pp. 174–75.

24. V. S. Clark, *History of Manufactures in the United States*, vol. 1, *1607–1860* (Washington, D. C., 1929), p. 261.

25. Ibid., p. 235.

26. Ibid., p. 418. See also note 65.

27. Ibid., pp. 418–22.

28. Musson, "Joseph Whitworth" (note 7).

29. In addition to the works by Roe, Woodbury, and Musson, previously cited, see L. T. C. Rolt, *Tools for the Job: A Short History of Machine Tools* (London, 1965); W. Steeds, *A History of Machine Tools* (Oxford, 1969); and S. B. Saul, "The Machine Tool Industry in Britain to 1914," *Business History* 10 (1968): 22–43.

30. Musson and Robinson, *Science and Technology* (note 7), pp. 23–24, 28, 50–51, 65, 81, 109–10, 132–33, 143, 433–40, and 455–58.

31. Ibid., chap. 13.

32. Roe, *English and American Tool Builders* (note 2), pp. 2–3.

33. *Select Committee on Exportation of Machinery*, Parliamentary Papers, 1841, vol. 7, *Second Report*, p. vii.

34. Evidence by William Jenkinson, a Salford engineer, in ibid.

35. J. Nasmyth, "Remarks on the Introduction of the Slide Principle in Tools and Machines employed in the Production of Machinery," in R. Buchanan, *Practical Essays on Mill Work* (3d ed., rev. by G. Rennie, 1841), pp. 393–418; W. Fairbairn, British Association *Report*, 1862, pp. lxiii–lxiv.

36. H. W. Dickenson, "Origin and Manufacture of Wood Screws," *Newcomen Society Transactions* 22 (1941–42): 79–89; Rolt, *Tools for the Job* (note 29), pp. 58–59; J. P. M. Pannell, "The Taylors of Southampton," Institution of Mechan-

ical Engineers, *Proceedings* 169 (1955): 924–31.

37. In addition to the general works on machine tools previously cited, see H. W. Dickenson, "Joseph Bramah and his Inventions," *Newcomen Society Transactions* 22 (1941–42): 169–86; Roe, "Interchangeable Manufacture" (note 2); I. McNeil, *Joseph Bramah* (London, 1968); P. Clements, *Marc Isambard Brunel* (London, 1970); J. F. Petree, *Henry Maudslay (1771–1831) and Maudslay, Sons & Field Ltd.* (London, 1949), and "Henry Maudslay—Pioneer of Precision," Institution of Mechanical Engineers, *Engineering Heritage* 1 (London, 1964); C. C. Maudslay, *Henry Maudslay* (London, 1948); K. R. Gilbert, *The Portsmouth Blockmaking Machinery* (London, 1956), and "Henry Maudslay 1771–1831," *Newcomen Society Transactions* 44 (1971–72): 49–62.

38. Farey gave this account during discussion of John Chubb's paper "On the Construction of Locks and Keys," Institution Civ. Eng., *Proceedings* 9 (1849–50): 331–32.

39. Roe, *English and American Tool Builders* (note 2), p. 17.

40. Ibid., pp. 5, 138–40. For a similar statement by Roe, see his article on "Interchangeable Manufacture" (note 2).

41. Habakkuk, *American and British Technology* (note 5), p. 120.

42. Roe, *English and American Tool Builders* (note 2), pp. 57–59; Rolt, *Tools for the Job* (note 29), pp. 92–103; Steeds, *A History of Machine Tools* (note 29), pp. 35–37, 40–42; Musson, "Joseph Whitworth" (note 7), pp. 111, 114–115. Whitworth worked at Maudslay's, Holtzappfel's, and Clement's before setting up his own workshop in Manchester.

43. E. S. Ferguson, *Early Engineering Reminiscences (1815–1840) of George Escol Sellers* (Washington, D. C., 1965), p. 127. For Donkin, see S. B. Donkin, "Bryan Donkin . . . 1768–1855," *Newcomen Society Transactions* 27 (1949–50 and 1950–51): 89–95.

44. Musson and Robinson, *Science and Technology* (note 7), chap. 14.

45. In addition to the references in Roe, Rolt, and Steeds, see Musson and Robinson (note 7), pp. 478–89; Musson, "Joseph Whitworth" (note 7), pp. 110–13, 123–25; and H. W. Dickenson, "Richard Roberts, His Life and Inventions," *Newcomen Society Transactions* 25 (1945–47): 123–37, as well as *The Engineer*, 14 Feb. and 17 Mar. 1947.

46. A. Ure, *Philosophy of Manufactures* (1835), p. 37. For further contemporary emphasis on Roberts's achievements in standardized manufacture, see *The Engineer*, 20 Feb. 1863, and Institution Civ. Eng., *Proceedings* 24 (1864): 537.

47. S. Smiles, *Industrial Biography* (London, 1863), p. 271. Roberts certainly preceded Whitworth in his use of standardized "plug-and-ring" and "go-no go" gauges in the 1820s, and they were, in fact, of earlier origin.

48. Institution Civ. Eng., *Proceedings* 28 (1868–69): 573–608; Roe, *English and American Tool Builders* (note 2), pp. 75–80; Rolt, *Tools for the Job* (note 29), chap. 6; D. Brownlie, "John George Bodmer: His Life and Work," *Newcomen*

Society Transactions 6 (1925–26): 86–110; Musson, "Joseph Whitworth" (note 7), pp. 113, 124–25, 130.

49. *The Engineer*, 20 Feb. 1863; Institution Civ. Eng., *Proceedings* 28 (1868–69): 589.

50. Institution Civ. Eng., *Proceedings* 28 (1868–69): 576.

51. A.E. Musson, "James Nasmyth and the Early Growth of Mechanical Engineering," *Economic History Review*, 2d ser. 10 (1957–58): 121–27. This has been reproduced in a revised form in Musson and Robinson (note 7), chap. 15, in which further references are given.

52. These references are from Nasmyth's letters in 1836 and from a description of his works in Love and Barton, *Manchester As It Is* (Manchester, 1839), pp. 213–19.

53. As emphasized in my original 1957 article.

54. *The Engineer*, 23 May 1856.

55. Saul, *Technological Change* (note 8), pp. 146–47.

56. These are all discussed in Musson, "Joseph Whitworth" (note 7). Like Maudslay, Whitworth was particularly interested in achieving precision engineering and standardization, based on accurate standard measures and gauges.

57. Institution of Mechanical Engineers, *Proceedings*, Feb. 1887, p. 153. There is a similar statement in *The Times* obituary of Whitworth, 24 Jan. 1887, where the idea of distributing the work "among the best machine shops in the country" is attributed to John Penn, the famous London engineer, but the credit is given mainly to Whitworth, whose "standards of measurement and of accuracy and finish were by that time thoroughly recognized and established throughout the country." See also G.A. Osborn, "The Crimean Gunboats," *The Mariner's Mirror* 51 (1965): 106–8, and Saul (note 8), p. 151, who points out the leading role played by Penn's and Maudslay's of London in this striking early example of mass-production marine engineering.

58. Musson, "Joseph Whitworth" (note 7), p. 120.

59. Habakkuk, *American and British Technology* (note 5), pp. 151–52 et passim. For a fuller discussion of Habakkuk's views, see Musson, "Joseph Whitworth" (note 7), pp. 129–35.

60. N. Rosenberg, "Anglo-American Wage Differentials in the 1820s," *Journal of Economic History* 27 (1967): 221–29.

61. Musson and Robinson (note 7), pp. 437–39, 485, 505–7; Musson, "Joseph Whitworth" (note 7), pp. 133–35.

62. Love and Barton, *Manchester As It Is* (note 52), pp. 210 ff.; Musson and Robinson (note 7), p. 485.

63. Habakkuk, *American and British Technology* (note 5), pp. 152–53; Musson and Robinson (note 7), chap. 15 passim; Musson, "Joseph Whitworth" (note 7), pp. 134–35.

64. Saul, *Technological Change* (note 8), chap. 5; R.C. Floud, "The Adolescence of American Engineering Competition, 1860–1900," *Economic History Review*, 2d ser. 27 (1974): 57–71.

65. Victor S. Clark was notable in emphasizing British superiority in machine tools and the limited extent of American developments in standardized manufacture at this time. See above, note 24.

66. *Report of the Select Committee on Small Arms*, Parliamentary Papers, 1854, XVIII, Q.2,043. Whitworth remarked on the failure to extend the use of woodworking machinery in England, after the remarkable success of Brunel and Maudslay with their blockmaking machinery (*New York Industrial Exhibition: Special Reports of Mr. George Wallis and Mr. Joseph Whitworth*, Parliamentary Papers, 1854, vol. 26, p. 116).

67. Whitworth's *Special Report* (note 66), p. 112.

68. *Report of the Committee on the Machinery of the U.S.A.*, Parliamentary Papers, 1854–55, vol. 50, p. 578.

69. *Small Arms Committee Report*, Q.1,367. See also *James Nasmyth, Engineer: An Autobiography*, ed. S. Smiles (1883), pp. 362–63.

70. Musson and Robinson (note 7), p. 507.

⚙⚙⚙⚙⚙⚙⚙⚙⚙⚙⚙⚙

WHY IN AMERICA?

Nathan Rosenberg

Let me try to clarify just what question it is I am trying to answer. I am not going to discuss the high overall rate of economic growth in the U.S. after 1800, i.e., the rate of increase of per capita income. In fact, that rate of growth was not particularly high by comparison with later experience, although many scholars believe it accelereated during the 1830s. If we extend our time horizon well back into the colonial period, we find that the rate of growth of per capita income between 1710 and 1840 did not exceed one-half of one percent per year.[1] Nor are we considering the overall rate of technological innovativeness which characterized the American economy, although that rate was doubtless very high. In spite of continued obeisance to the idea of Yankee ingenuity, it cannot be overstressed that America in the first half of the nineteenth century was still primarily a borrower of European technology. Although the rate of technical change was indeed high, most of the new technologies were not American inventions. Americans were rapid adopters of foreign technologies when it suited their economic needs, and they also were skillful in modifying someone else's technology to make it more suitable to their needs. They made abundant albeit selective use of European innovations in power generation, transportation, and metallurgy.[2] It is worth noting, moreover, that high rates of technological improvement were not always a part of the American experience. Such improvements seem to have played an insignificant role in the American economy in the eighteenth century. The latest scholarship on the colonial period suggests that the rate of inventive activity was very low, and that the slow but steady growth in productivity was dependent upon a cumulatively powerful combination of forces centering upon improvements in the organization of industries and the more effective functioning of markets and ancillary institutions.[3]

The question I propose to address, then, is not the high rate of technological change in America when that rate indeed became high in the nineteenth century. Rather, the question is one of direction and character. By the time of the Crystal Palace Exhibition in London in 1851, there was something sufficiently distinctive about many American goods that the British coined the phrase "the American system of manufactures." The expression was used to describe goods which were (1) produced by specialized machines, (2) highly standardized, and (3) made up of interchangeable component parts.[4] Other contributors will examine the American system of manufactures more carefully. I propose to address myself to the question of causality. Why did this system emerge first in America? By midcentury it was apparent that there was an important class of goods which was being produced in a distinctive way in America.[5] It is important, however, that the distinctiveness should not be overstated, nor should it be thought that America also had assumed a role of broad technological leadership across the whole manufacturing sector. That was far from the case. Indeed, even in gunmaking, commonly regarded as the main triumph of the American system of manufactures, American barrelmaking may still have lagged well behind Britain at midcentury.[6]

Now, if you call upon an economist to try to explain a particular phenomenon, such as why the technology which we have come to call the American system of manufactures first arose in the United States and not elsewhere, you are not entitled to express surprise over an explanation couched in terms of supply-and-demand analysis. That is what I propose to deliver, and I hasten to add that I do so without apology. Presenting a historical explanation in terms of supply-and-demand forces does not *necessarily* mean that I am committed to some form of economic determinism. Supply and demand are nothing more than convenient conceptual categories which enable us to think about an event or process in a more systematic way and to organize our analysis in a way which lends itself more readily to an understanding of cause-effect relationships. The forces *underlying* these economic categories may be social, geographic, technological, or ideological, but such forces have economic content and effects, and I am particularly interested in focusing upon the economic implications or consequences of these forces. There is a difference, which I trust that I need not belabor, between *translating* certain phenomena into their economic consequences and insisting that economic variables are all powerful.

I have already suggested a distinction between the rate and direction of technological progress. However, these two factors are not so easily separable as may be thought at first glance. The point is that differences in the resource endowment and demand conditions of an economy go a long way toward determining what kinds of inventions—with what kinds of product characteristics and factor-saving biases—it will be profitable to develop and exploit. To the extent that technological progress is responsive to economic forces—and I would agree that it is highly responsive and needs to be understood in these terms—the inventions actually brought forth in a particular country will tend to be compatible with those special needs. We need to distinguish here between invention and adoption. It is obvious that only those inventions which are compatible with a country's needs will be widely adopted. I am making here the stronger assertion that a high proportion of the *inventions made* will also reflect the peculiar needs of the economic environment in which they are developed.

If, at any given time and state of technical knowledge, certain *kinds* of inventions are easier to create, for whatever reason, economies where such inventions are also economically appropriate would be expected (*ceteris paribus*, as economists are fond of saying) to find it easier to generate new and improved technologies. If economic forces tend to push economies with different factor endowments in different directions in the attainment of technological progress, and if the technical problems which have to be overcome are more difficult in some directions than others, then the rate of technological progress may not be independent of the direction in which the economy is being pushed.

I have developed this argument because I believe there is something to it. The American experience in the nineteenth century involved an economy with resource endowment and demand conditions that pushed it in a direction somewhat different from the one that prevailed at the time for the economies of western Europe. This turned out to be a direction in which the inventive payoff was rather high. You may, if you wish, take this as a confirmation of the belief that a benevolent Providence watched over the affairs of America in the nineteenth century. Or you may simply regard it as luck or as the consequence of a particular resource endowment.

Whatever the label, I want to argue that what we call the American system of manufactures was a part of a larger process of economic adaptation, or even, if you prefer, economic evolution. American economic and social conditions in the nineteenth century gave rise to a certain direction of thrust in the exploration for new techniques. It was a direction which, both for social and economic reasons, was less appropriate in Europe. And, as it turned out, it was a direction in which there happened to exist a rich layer of inventive possibilities. In this sense, I will argue, the American system of manufactures was a species of a larger genus and, in this sense also, the high *rate* of technical progress was not entirely independent of the *direction* which that progress took.

The notion that the American system of manufactures requires explanation is reinforced by the facts that (1) the United States was, in the first half of the nineteenth century, still largely a technical borrower rather than a pioneer, and (2) technical innovations incorporating some of the elements of the system had been introduced earlier in Britain, mainly in the Portsmouth naval dockyards, but were not widely adopted elsewhere in that country.

Demand

What were the factors on the demand side that were so conducive in early-nineteenth-century America to the emergence of the American system of manufactures? There was a very rapid rate of growth in the aggregate demand for certain classes of commodities, which encouraged entrepreneurs to devise or adopt new production processes whose profitability required very long production runs. Adam Smith long ago taught that "the division of labor is limited by the extent of the market." The proposition applies to the use of specialized machinery as well as to specialized men. The fixed costs embodied in special-purpose machinery are warranted only if they can be distributed over a large volume of output. Therefore the confident expectation of rapidly growing markets is an extremely powerful inducement. Furthermore, the use of highly specialized machinery, as opposed to machinery which has a greater general-purpose capability, is contingent upon expectations concerning the *composition* of demand—specifically, that there will be strict and well-defined limits to permissible variations. Both of these conditions—rapid growth in demand and circumstances conducive to a high degree of product standardization—were amply fulfilled in early-nineteenth-century America.

What forces were responsible for these conditions? Probably the most pervasive force of all was the extremely rapid rate of population growth, primarily from natural increase but with immigration assuming a role of some significance in the 1840s. Between 1790 and 1860, the American population grew at a rate of nearly three percent per year (immigration included), a rate more than twice as high as that achieved by any European country even for much shorter periods. American fertility levels at the beginning of the nineteenth century were close to fifty per thousand—again, a rate far higher than European levels—and crude death rates were somewhere around twenty per thousand. The factors underlying this rapid expansion in population are many, but most important was probably the abundant supply of high-quality land in a society which was still predominantly agricultural. The abundance of good land, and the concomitant optimism about the future, were conducive to high fertility levels as well as an inducement to immigration.[7]

Although scholars differ concerning the long-term trend in the rate of growth of per capita income during our period, there can be no doubt that that strong rate of population growth resulted in an exceedingly rapid rate of market growth. As early as 1840 the net national product of the United States was two-thirds or more the size of Great Britain's.[8] In subsequent years America's net national product grew far more rapidly than Britain's.

Rapid population growth resulted in a very high rate of new household formation and therefore a rapid rate of growth in the demand for a wide range of manufactured commodities. This growth in market size was reinforced by other developments. Improvements in transportation, especially after the beginning of the canal-building period in 1815 and the beginning of railroad construction in the 1830s, served to link industrial centers with remote potential markets that previously had been largely self-sufficient. Thus, in many cases industrial producers and consumers were in a position to establish economically feasible market relationships with one another for the first time.

Many of the distinctive characteristics of the American market can be properly appreciated only in the context of its predominantly rural and agricultural character. Over eighty percent of the American labor force was in agriculture in 1810 and this remained true for well over sixty percent in 1840.[9] The predominance of the agricultural sector in the nineteenth century, and the abundant endowment of natural resources more generally, served to shape the American environment in numerous ways which made it more receptive to the American system of manufactures. First of all, we have already noted how the rich and abundant supply of agricultural land contributed to high fertility levels and therefore to a rapid growth in the market. In addition, the abundance of land conferred another advantage upon the American economy, for it meant that food prices were relatively low. As a result, for any given income level or family size, there was a larger margin left over for the purchase of nonfood products, including those produced under the American system of manufactures. Thus the structure of relative prices in the United States, as a result of its resource endowment, was distinctly more favorable than in Europe for the emergence of manufactured goods. Both American farmers and urban industrial workers spent less of their income on food than did their European counterparts. Empirical evidence in support of this hypothesis has been carefully marshalled and analyzed by Albert Fishlow, relying upon data collected for the period 1888–91.[10] Since American food prices were, relatively, lower earlier in the nineteenth century, the case for the more favorable American conditions is even stronger.[11]

Yet another advantage conferred by America's natural-resource abundance was the easy accessibility to land ownership and the relatively egalitarian social structure which consequently emerged—with the major exception, of course, of the southern plantation system. Outside of the south there had emerged in colonial times and later a rural society which was far less hierarchical, and with property ownership far less concentrated, than was the case in Europe. American society included a much larger component of middle-class farmers and craftsmen. This was in sharp contrast with European societies in which poor peasants and farm laborers, with very little income beyond subsistence needs, constituted a large fraction of the total population, while at the top of the social pyramid was a small group of large landowners whose expenditure patterns made little contribution to the emergence of a sector producing standardized manufactured goods.

In America, to a degree which is now largely lost from the collective memory of a highly urbanized population, markets in the first half of the nineteenth century were dominated by the tastes and requirements of middle-class rural households.[12] In such households, often living under frontier or at least isolated conditions, there was little concern—indeed there was little opportunity—for ostentation or conspicuous consumption. Out of these social conditions emerged a relative simplicity of taste and a stress upon functionalism in design and structure. Rural isolation also strongly favored reliability of performance and ease of repair in case of breakdown. Such characteristics were particularly important for agricultural machinery and firearms. Simplicity of design and uniformity of components meant that a farmer could repair a broken plough in the field himself without having to depend upon a skilled repairman or distant repair shop. Thus, out of the social and geographic conditions of land-abundant America emerged a set of tastes and preferences highly congenial to a technology capable of producing large quantities of standardized, low-priced goods. These circumstances even left their indelible imprint on the American automobile in the early years of the twentieth century. The Ford Model T was designed in a manner which strongly resembled the horse and buggy, and the primary buyers were farmers for whom a cheap car offered a unique opportunity for overcoming rural isolation.

Supply

The point has already been made that in the early nineteenth century America was a large-scale borrower of European technology. But the point also has been made that America's factor proportions, especially its rich abundance of natural resources, differed substantially from Europe's.

These two propositions have an interesting implication that provides a useful way of entering into an examination of supply-side influences upon the emergence of the American system of manufactures. For, if factor proportions between the two continents were sufficiently disparate, it would follow that the technology devised to accommodate European needs did not always constitute, in present-day terminology, an "appropriate technology" in America. We now need to consider, therefore, how differences in the supply of available resources pushed Americans in a direction that helps account for the country's unique technological contributions.

An abundance of natural resources means, in economic terms, that it is rational to employ methods of production which are resource intensive. Much of American inventive activity in the first half of the nineteenth century aimed at substituting abundant natural resources for scarcer labor and capital—although this aim is often not apparent from a present-day examination of the hardware devised for the purpose. America's early world leadership in the development of specialized woodworking machinery—machines for sawing, planing, mortising, tenoning, shaping, and boring—was a consequence of an immense abundance of forest products. Although these machines were wasteful of wood, that was of little consequence in a country where wood was very cheap. The substitution of abundant wood for scarce labor was, in fact, highly rational.[13]

Similarly, the immense available supply of potential farmland was not conducive to inventive activity that would maximize output per acre (as would be sensible in a land-scarce economy like Japan's), but rather to activity that would maximize the amount of land that could be cultivated by a single worker. This is precisely the thrust of nineteenth-century American mechanical innovation in agriculture, based upon animal power which supplied the traction for the steel plough (as well as Jethro Wood's earlier plough which already had replaceable cast-iron parts); the cultivator which replaced the hand-operated hoe in the corn and cotton fields; and the awesome reaper which swept away what once had been a basic constraint upon grain cultivation—fluctuations in labor requirements that reached a sharp peak during the brief harvesting season. Later, output per worker was further increased by binders and threshing machines and, eventually, by combine harvesters. In corn cultivation, the corn sheller and the corn picker were particularly valuable for their laborsaving characteristics.

In the gunmaking trade, usually regarded as the *locus classicus* of the distinctly American mass production technology, some of the most original contributions were in the development and elaboration of a set of lathes for shaping the gunstock with a minimum amount of labor. The fact

Broadside (1852), from one of the first major manufacturers of woodworking machinery in the U.S.A. (From the collections of the Division of Mechanical and Civil Engineering, National Museum of American History; Smithsonian Institution Neg. No. 81–190.)

Thomas Blanchard's gunstock-duplicating lathe, built for the Springfield Armory in 1822. (Smithsonian Institution Neg. No. 24437.)

that these laborsaving machines were initially highly wasteful of wood—as compared to the shaping of a gunstock by a Birmingham craftsman—was a trivial consideration in a wood-abundant economy. Blanchard lathes were widely adopted in America.

What is the relevance of these woodworking and agricultural machines to the emergence, by midcentury, of the American system of manufactures? The essential point is that resource abundance provided an incentive in America to explore the possibilities of certain new machine technologies earlier and more deeply than was the case in Europe.[14] The fact that these machines, particularly in their early stages of development, were not only laborsaving but also resource intensive, was not the economic deterrent that it was in Britain or France. Thus, in a wide range of manufacturing activities, Americans had a strong incentive to develop and adopt new technologies which, in effect, traded off abundant natural-resource inputs for labor, as well as machines which were wasteful of natural resources yet could be constructed more cheaply. But the pressures generated by America's unique resource endowment led to exploratory activities and to eventual learning experiences the outcome of which cannot be adequately summarized merely in terms of factor-saving or factor-using biases. For they also led to new patterns of specialization and division of labor between firms—especially between the producers and users of capital goods—as a result of which the American economy developed a degree of technological dynamism and creativity greater than existed in other industrial economies in the second half of the nineteenth century.

I believe that this technological dynamism was due in large measure to the unique role played by the capital-goods industries in the American industrialization process and the especially favorable conditions under which they operated. For these capital-goods industries—I refer here primarily to those involved in the forming and shaping of metals—became learning centers where metalworking skills were acquired and developed, and from which such skills were eventually transferred to the production of a sequence of new standardized products—interchangeable firearms, clocks and watches, agricultural machinery, hardware, sewing machines, typewriters, office machinery, bicycles, automobiles. A key feature of the story of American industrialization is that it involved the application of certain basically similar production techniques to a growing range of manufactures. Moreover, the technological knowledge and competence that gradually accumulated in this sector were directly applicable to generating cost reduction in the production of capital goods themselves. A newly designed turret lathe or universal milling machine, or a new steel alloy permitting a lathe to remove metal at higher speeds—each of these innovations not only resulted in better machines but also reduced the cost of producing the machines in the first place.

Thus, although the initial shift to the capital-using end of the spectrum was generated by the unique pattern of American resource scarcities, it is by no means obvious that the final outcome of this process was in terms of factor biases. For the capital-using path was also a path which, *eventually*, generated a much-increased capacity for capital-saving innovations. The attempt to deal with labor scarcity in a regime of natural-resource abundance pushed us quickly in a direction in which there turned out to be rich inventive possibilities. In turn, the skills acquired in a more capital-abundant society with an effectively organized capital-goods sector provided the basis—in terms of knowledge and engineering skills and expertise—for innovations which were capital saving as well as labor saving. Indeed, most new products, after their technical characteristics became sufficiently stabilized, have passed through such a cost-reducing stage during which capital-goods producers accommodated themselves more efficiently to the large-quantity production of the new product. American industry seems to have particularly excelled at these activities.

Aside from the highly visible major inventions, capital-intensive technologies have routinely offered extensive opportunities for improvements in productivity which seem to have had no equivalent at the labor-intensive end of the spectrum. Knowledge of mechanical engineering, metallurgy, and, perhaps most important of all, the kind of knowledge which comes from day-to-day contact with machine technology, provide innumerable opportunities for small improvements—minor modifications, adaptation to some special-purpose use, design alterations, substitution of a superior or cheaper material—the cumulative effects of which have, historically, been very great.

It is important that these developments be seen in their actual historical sequence. America began the growth of her capital-goods sector not only with a strong preoccupation with standardization and interchangeability, but also with some early experience with techniques as well as a market which readily accepted standardized products. This shaped the eventual outcome of the industrialization process in some decisive ways. The acceptance of standardization and interchangeability vastly simplified the production problems confronting the makers of machinery and provided the technical basis for cost reductions in machinemaking. At the same time it provided the conditions which encouraged the emergence of highly specialized machine producers as well as the transfer of specialized technical skills from one industrial application to another. Indeed, America's most significant contributions to machine-tool design and operation and related processes—profile lathes, turret lathes, milling machines, die-forging techniques, drilling and filing jigs, taps and gauges—were associated with specialized, high-speed machinery devoted to the production of standardized components of complex products.

Thus, as a result of certain initial conditions, America's industrialization in the early nineteenth century proceeded along one out of a variety of possible paths. It was a path dictated by peculiar resource conditions and also a path rich in inventive possibilities. Movement along this path generated a dynamic interaction of forces, a dialectical process in which cause and effect become exceedingly difficult to disentangle. While greater homogeneity of tastes was originally conducive to the introduction of goods produced according to the American system of manufactures, it is also true that, once this technology began to spread, it in turn shaped and influenced tastes in the direction of simplicity and functionality.[15] Furthermore, although American factor endowment pushed in the direction of mechanization, the experience with mechanization *in itself* brought about an improvement in inventive ability and its more rapid diffusion. This was an interactive process, whereby the successful application of mechanical skills in one sector improved the likelihood of their successful application in other sectors.

What, finally, can we add by way of explaining why Britain—which was far more advanced industrially than the United States in the early nineteenth century—did not go further than it did in developing the distinctive features of the American system of manufactures? Much of the answer is already implicit in the preceding discussion of Anglo-American differences in conditions of demand-and-resource endowment. There is, however, an additional factor which deserves mention, namely, the persistence and continuing powerful influence of certain traditional values and attitudes which had their roots in a preindustrial craft society. Such values and attitudes had, of course, played a much more significant role in shaping the outlook of workers and engineers in Britain than in the United States. Strong craft traditions, with their emphasis upon pride in workmanship, individuality, and high standards of product quality, were often inimical to standardization and the alterations in final product design which were essential to low-cost, high-volume production. To some extent the persistence of traditional craft attitudes in Britain reflected the demand-side phenomena to which we have already referred, as well as other market peculiarities. But beyond that there is evidence from many industries—firearms, automobiles, locomotives, a wide range of production machinery, clocks and watches—suggesting a preoccupation with technical perfection beyond what could be justified by narrowly utilitarian considerations.[16] And perhaps this observation suggests what should be an appropriate final comment upon why this particular system first emerged in America. For the American system of manufactures was, above all, a totally unsentimental approach to the productive process in industry, one in which purely commercial considerations prevailed. The British long consoled themselves with the belief that standardization and mass-production techniques inevitably resulted in an inferior product. Unhappily, from their point of view, they persisted in this belief long after it ceased to contain even an element of truth.

Notes

1. Robert Gallman, "The Record of American Economic Growth," in Lance Davis et al., *American Economic Growth* (New York, 1972), p. 22.

2. Nathan Rosenberg, *Technology and American Economic Growth* (New York, 1972), chap. 3.

3. See James Shepherd and Gary Walton, *Shipping, Maritime Trade and the Economic Development of Colonial North America* (Cambridge, Mass., 1972).

4. The degree or ease of interchangeability varied between products and over time. The degree of precision required for interchangeability varied considerably from one product to another. As one nineteenth-century commentator observed: "The more prevalent modern idea of the interchangeable in mechanism supposes a super-refinement of accuracy of outline and general proportions that is not always necessary or even desirable. It would be a criminal waste of time and substance to fit a harrow tooth with mathematical accuracy, but yet any harrow tooth should have a practical interchangeable relation to all harrows for which it is designed. The instructive rewards of folly would certainly overtake him who should attempt to make ploughshares and coulters with radical exactness; nevertheless, these essential parts of ploughs should be interchangeable among all ploughs to which they are adapted." W. F. Durfee, "The History of Interchangeable Construction," A.S.M.E. *Transactions* 14 (1893): 1228.

5. The role of the manufacturing sector had increased very sharply during the decade of the 1840s. According to Gallman's estimates the contribution of manufacturing to national commodity output grew from 17% to 30% during the decade 1839–49. Robert Gallman, "Commodity Output, 1839–1899," in the Conference on Income and Wealth's *Trends in the American Economy in the 19th Century* (Princeton, 1960), p. 26.

6. Paul Uselding, "Henry Burden and the Question of Anglo-American Technological Transfer in the Nineteenth Century," *Journal of Economic History* 30 (1970): 312–37.

7. For an excellent, concise treatment of long-term demographic trends in America, see Richard A. Easterlin, "The American Population," chap. 5 in Lance Davis et al., *American Economic Growth* (note 1).

8. Gallman, "Commodity Output, 1839–1899" (note 5), p. 33.

9. Lebergott's estimates of the percentage of the labor force in agriculture are

as follows: 1810—80.9%; 1820—78.8%; 1830—68.8%; 1840—63.1%; 1850—54.8%; 1860—52.9%; 1870—52.5%; 1880—51.3%. Stanley Lebergott, "Labor Force and Employment, 1800–1960," in *Output, Employment and Productivity in the U.S. after 1800*, ed. Dorothy Brady, Studies in Income and Wealth, vol. 30 (National Bureau of Economic Research, 1966), p. 119.

10. Albert Fishlow, "Comparative Consumption Patterns, The Extent of the Market, and Alternative Development Strategies," in *Micro-Aspects of Development*, ed. Eliezer Ayal (New York, 1973).

11. "The very fact that food prices should have been relatively lower earlier would have given them more leverage at precisely the right moment—when per capita incomes were smaller and the percentages allocated to food higher. What is significant is that they did not rise rapidly thereafter and impede the extension of the market. American food prices in the 1830s and 1840s were not only lower relative to those of a nonfood composite, but also compared to British foodstuffs. That is, British food prices fell more rapidly from 1831–50 to 1866–90 than did American, and we have already seen that American foods were absolutely cheaper at that later date. Hence, the comparative advantage the United States enjoyed was greater in the earlier period." Ibid., pp. 77–78.

12. "The share in aggregate expenditures of different groups in the population in the nineteenth century and earlier can be estimated only roughly, for want of statistical information representative of all sections of the country. Even a crude statistical picture, however, can indicate how important rural households were in the market for manufactured goods and services. In the 1830s, about 66 percent of all households were on farms, 23 percent in villages, and 11 percent in cities. On the supposition that farm families on the average were able to spend $100, village families $200, and city families $300 over and above their outlays for housing and fuel, rural families (farm and village) would have accounted for 76 percent of all such expenditures." D. Brady in Davis et al., *American Economic Growth* (note 1), p. 62.

13. For a more detailed treatment, see Nathan Rosenberg, *Perspectives on Technology* (Cambridge, 1976), chap. 2, "America's Rise to Woodworking Leadership."

14. The rest of this paragraph draws upon Nathan Rosenberg, "American Technology: Imported or Indigenous?" *American Economic Review Papers and Proceedings*, Feb. 1977.

15. Dorothy Brady, "Relative Prices in the Nineteenth Century," *Journal of Economic History* 24 (1964): 147–48.

16. S.B. Saul, "The Market and the Development of the Mechanical Engineering Industries in Britain, 1860–1914," *Economic History Review* 41 (1967); S.B. Saul, "The Engineering Industry," chap. 7 in *The Development of British Industry and Foreign Competition, 1875–1914*, ed. D. Aldcroft (Toronto, 1968); S.B. Saul, "The Motor Industry in Britain to 1914," *Business History* 5 (1962); R.A. Church, "Nineteenth Century Clock Technology in Britain, the U.S., and Switzerland," *Economic History Review* 49 (1975); and Nathan Rosenberg, intro. to *The American System of Manufactures* (Edinburgh, 1969).

MILITARY ENTREPRENEURSHIP

Merritt Roe Smith

Government's pivotal role in the antebellum American economy is well known and thoroughly documented.[1] Indeed, all levels of government—federal, state, and local—evolved policies aimed at insuring domestic stability and promoting economic growth. The magnitude of this involvement ranged widely from direct grants and subsidies to such indirect aids as the provision of banking facilities, tax exemptions, and tariff and patent legislation. Confronted by barriers to expansion, nineteenth-century Americans expected public institutions to enter the economic arena by assuming initial risks, removing bottlenecks, and cultivating resources until private enterprises could become established on a profitable basis. Besides fostering a climate in which business could organize and flourish, the actual presence of government agents in developing regions had an incalculable psychological impact which spurred public confidence, attracted investment, and stimulated industrial development. Such activities not only served utilitarian purposes but also strengthened popular beliefs in progress, prosperity, and perfectability. On these ideological foundations rested the viability of republican institutions and the promise of American life.

One of the most visible agents of government enterprise was the United States Army. Due partly to a traditional commitment to national security, partly to a romantic attachment to Manifest Destiny, and partly to an abiding concern for conveying a favorable image in a society inherently skeptical of standing armies, members of the military willingly, even eagerly, participated in programs for internal domestic improvement. While the regular Army and Quartermaster Department contributed significantly to the settlement and stabilization of the frontier, the real thrust of military enterprise emanated from three technical bureaus housed in the War Department: the Corps of Engineers; the short-lived Topographical Bureau (1838–63); and the Ordnance Department. These bureaus undertook numerous projects directly affecting the course of diplomacy, science, and industry in America. Precisely because their activities touched so many vital interests in the republic, they occupied a central position in programs for national development. Such intimate involvement in the affairs of state allowed them to demonstrate their usefulness to society and even made military life exciting. Yet in the highly partisan ferment of Jacksonian America, not everything they did with internal improvements proved satisfying. Since jobs, contracts, and general regional prosperity weighed in the balance, the bureaus could not escape controversy. Indeed, as public agencies charged with civil as well as military duties, they frequently found themselves in the anomalous position of having to serve rival constituencies with all the associated pressures of congressional scrutiny, sectional jealousy, and pork-barrel politics. That they succeeded in addressing problems of national growth in a tempestuous era testifies not only to their professional zeal but also to their role as catalysts of change and consolidation.

Much has been written about the Topographical Bureau and the Corps of Engineers, whose extensive explorations, geodetic surveys, and construction activities resulted in an impressive fund of scientific data as well as in a wide variety of civil works.[2] A good deal less is known about the exploits of the Ordnance Department, especially its involvement in one of the great technological achievements of the nineteenth century popularly known as the "American system of manufactures." The term evidently originated in 1854 when a British military commission investigated the machinery of the United States. In their published report, the commissioners mentioned "the American system" with specific reference to the division of labor and application of machinery in the production of firearms with interchangeable parts. That the expression was printed in lower case and appeared only once in the report did not seem to bother those historians who have seized upon it as a catch-all phrase to describe developments in the American small-arms industry prior to the Civil War.[3] Interestingly enough, the ordnance officers and civilian mechanics who developed the system never used the phrase, preferring instead to describe their work as part of a larger program to "insure system and uniformity" in all materiel of war.[4] Since the subject occupied the minds and efforts of ordnance officials for more than forty years, an evaluation of their contribution is needed. Therefore, this essay seeks to analyze the evolving concept of uniformity and to assess the Ordnance Department's role in facilitating the rise of the new technology.

The year 1815 was one of jubilation in the United States. The War of 1812 had ended and, amidst bonfires, cannon salutes, and tolling church bells, citizens everywhere joined in a national celebration the likes of which had not been seen since the Revolution. Republicanism had been vindicated, despotism vanquished, and independence finally achieved. Exhilarated by the public optimism and air of self-congratulation that followed the Treaty of Ghent, the popular press conveyed the feeling that a new era of "Peace and Plenty" was at hand. Slowly but perceptibly the Revolutionary ideals of liberty and republican virtue were beginning to find renewed meaning in expectations of progress and prosperity. In such a milieu, nothing seemed impossible. Wiser men knew, however, that the country had barely skirted disaster during the war; good fortune rather than military might had ensured the nation's survival. From the beginning, faulty arms, insufficient supplies, and tactical errors had plagued the war effort—facts that no one, in 1815, appreciated more than the secretary of war and his general staff.[5]

Three years earlier, at the urging of the executive branch, Congress had attempted to revamp the cumbrous army supply system by creating three central bureaus—the Quartermaster Department, the Commissary General of Purchases, and the Ordnance Department—to assist the beleaguered War Department in procuring, inspecting, and distributing military equipment. Because the legislation failed either to clarify the relationship between these agencies or to define adequately their respective duties, their activities overlapped at many points during the ensuing war years. These ambiguities resulted in a confusion of authority that perpetuated problems which had plagued the army for decades.[6]

With the end of hostilities Congress again sought to remedy the situation by passing, in February 1815, "An Act for the better regulation of the Ordnance Department." In contrast to the bill that had created this bureau in 1812, the new legislation carefully spelled out the duties and responsibilities of the office and expanded its sphere of authority. Formerly the department's primary mission had consisted of inspecting cannon, proving gunpowder, and supervising the manufacture and storage of gun carriages, munitions, and other equipment at several federal arsenals. Ordnance officers possessed no authority to make contracts, nor did they exercise jurisdiction over the procurement and production of small arms. Under the act of 1815, however, all this changed. In addition to transferring responsibility for the negotiation and supervision of all arms contracts from the Commissary General of Purchases to the Ordnance Department, the bill also placed the national armories at Springfield, Massachusetts, and Harpers Ferry, Virginia, under the latter's immediate command. Equally significant, the reorganization act empowered the chief of ordnance "to draw up a system of regulations... for the uniformity of manufactures of all arms ordnance, ordnance stores, implements, and apparatus, and for the repairing and better preservation of the same." For the next forty years this charge became the guiding principle of ordnance policy. Although little noticed at the time, the proviso set the stage for important developments in military technology and the eventual transformation of the American industrial system.[7]

Inquiry into the origins of the uniformity system sheds much light on the processes of technological innovation, but it also raises a fundamental question about the originality of the idea as an American invention. The United States was a developing nation in the early nineteenth century and, as such, relied on Europe both for manufactured goods and manufacturing techniques.[8] While the immediate need for uniform ordnance sprang from the bitter logistical experience of the War of 1812, the actual formation of the policy emanated from sources deeply rooted in the French military tradition. Since the Revolutionary War, French artillerists and engineers had exercised a pervasive influence on the United States Army. Through them, engineering treatises, testing procedures, arms designs, and educational techniques had made their way to American shores and were assimilated by native officers. Indeed, whenever the United States needed to revise and improve its military program, it looked primarily to the "French system" for appropriate models. This practice continued well into the 1840s.[9]

Among the most influential agents of French technology was Major Louis de Tousard. A graduate of the artillery school at Strasbourg, an aide to Lafayette during the American Revolution, and a skilled practitioner of military engineering, Tousard returned to the United States in 1793 after being convicted of subversive activities by the French National Convention and briefly jailed at the "bloody prison of l'Abbaye." Two years later he joined the newly created Corps of Artillerists and Engineers. He spent the next seven years on special assignments as an inspector of artillery and fortifications. Among other duties, he supervised the construction of several forts along the eastern seaboard, advised founders on the casting and boring of cannon, tested and experimented with iron ordnance, and initiated efforts to convert the garrison at West Point into a full-fledged military academy with a curriculum modeled after the École Polytechnique.[10] At President Washington's request he also initiated research on a three-volume work entitled *Amerian Artillerist's Companion*. This study, which was published in 1809, synthesized existing knowledge in the field and served as a standard text for American officers for more than a decade. Throughout his treatise Tousard emphasized the need to devise "a system of uniformity and regularity" based on scientific theory and experimentation. "This want of uniformity," he cautioned readers, "impeded for a long time the progress of the French artillery, as it will that of America, unless a similar system is adopted."[11]

The "system" to which Tousard referred was the brainchild of General

Jean-Baptiste de Gribeauval who in 1765 had initiated thoroughgoing reforms in the organization and production of French ordnance. Under Gribeauval's direction armorers had developed methods of manufacturing gun carriages and other equipment with standardized parts. Consolidated models, improved techniques, and rigid inspections characterized the new system. Its strategic significance was soon evidenced in the mobility of Rochambeau's light artillery during the American Revolution and later in Napoleon's speedy conquest of Italy.

British observers readily appreciated the tactical advantages of Gribeauval's uniformity of designs. "At the formation of their system," one officer wrote, "they saw the necessity of the most exact correspondence in the most minute particulars, and so rigidly have they adhered to this principle that, though they have several arsenals, where carriages and other military machines are constructed, the different parts of a carriage may be collected from these several arsenals, in the opposite extremities of the country, and will as well unite and form a carriage as if they were all made and fitted in the same workshop. As long as every man who fancies he has made an improvement is permitted to introduce it into our service," the officer concluded, "this cannot be the case with us."[12]

These and other reports prompted the British government to introduce similar changes in carriage construction, an undertaking that culminated around 1794 with the famous "stock-trail" arrangement featuring interchangeable wheels, pintle hooks, and wrought-iron axles. These improvements were eventually adopted by France in 1827 and, through French contracts, subsequently influenced American artillery designs during the 1830s. Yet, even though the stock-trail system eventually displaced Gribeauval's original artillery plan, his conceptual insights regarding "system and uniformity" made a deep impression on American officers and continued to provide an essential focus for their activities during the early national period.[13]

Gribeauval's influence also manifested itself in another way. As Inspector General of Artillery he had played an instrumental role in promoting the work of Honoré Blanc, a talented armorer who had developed various labor-intensive methods for manufacturing muskets with uniform parts. Favored by Gribeauval's patronage, Blanc had received approval to undertake experiments with the design and manufacture of small arms at several government installations. By the mid-1780s he had tooled the Vincennes arsenal with novel die-forging, jig-filing, and hollow-milling techniques capable of effecting "the greatest economy and the most exact precision."[14] These innovations had particular relevance for American manufactures. Thomas Jefferson, while serving as ambassador to Versailles, had visited Blanc in 1785 to witness the way in which the lock components of Blanc's muskets could be randomly selected and interchanged without any fitting or filing. Duly impressed by the demonstration, Jefferson wrote home about the experience and tried to persuade Blanc to emigrate to the United States. Although the Frenchman declined, Jefferson continued to monitor his activities and even shipped six of Blanc's muskets to Philadelphia in 1789. One person who profited from these diplomatic exchanges was Eli Whitney. Through correspondence and conversation with Jefferson, Secretary of the Treasury Oliver Wolcott, and other government officials, Whitney evidently learned of Blanc's work and tried to emulate it in filling his musket contracts with the War Department. Although his efforts fell short of success, he nonetheless became a zealous advocate of the uniformity principle, popularized the concept, and persuaded many politicans to support policies aimed at standardizing the manufacture of military arms.[15]

All these factors weighed upon the first chief of ordnance, Colonel Decius Wadsworth. As an intimate friend of Whitney's, he especially appreciated the importance of uniformity in small arms and had continually championed the New Haven manufacturer's interests at the War Department. As a former member of the Corps of Artillerists and Engineers, he had also worked closely with Tousard, Stephen Rochefontaine, and other Frenchmen in American service. Through them he became well acquainted with French practice, particularly Gribeauval's uniform system of artillery. Hence, upon assuming command of the Ordnance Department in 1812, he openly espoused *"Uniformity Simplicity* and *Solidarity"* and made the phrase his motto. The words poignantly reflected the degree to which he was influenced by French ideas and thereafter they shaped not only his actions but also those of his staff.[16]

As early as 1813, Wadsworth had expressed dissatisfaction with the hectic disarray of American ordnance. In language reminiscent of Tousard, he had written a fellow officer about the innumerable variations in artillery and of the urgent need to rectify the situation. In the past, he explained, "every superintendent selected whatever pattern and introduced whatever alteration his fancy suggested." Under present circumstances this could no longer be tolerated.

The necessity of some regulation to secure *simplicity* and uniformity must be obvious to all; yet men of reflection and experience alone can duly estimate the importance of these two qualities. Every variation in the proportions of pieces of the same calibre exacts a corresponding change in the carriage, and for every distinct calibre will be required not only a suitable carriage but its appropriate equipments and ammunition.

"In a word," he concluded, "unless the number of our calibres and their variations be reasonably reduced, and the whole be settled by some permanent regulation, no possible exertion can give to our artillery that perfection its importance merits and which the public service requires."[17]

Since similar deficiencies existed in the design and manufacture of small arms, Wadsworth called for a comprehensive reform, arguing that all ordnance should be consolidated, systematized, and rigidly regulated. Who would implement these changes? His answer, of course, was "the Ordnance Department." To this end he had actively lobbied to secure passage of the act reorganizing the department in February 1815. The language of the bill reflected his influence, and its comprehensiveness, as indicated earlier, provided him with the necessary authority to pursue his cherished goal of "system and uniformity."[18]

As a bachelor with few family obligations, Wadsworth completely immersed himself in professional affairs and made uniformity his special calling. Whether scrutinizing the propositions of private contractors, enforcing departmental discipline, or reorganizing the federal arsenals into those for "construction" and those for "storage and repair," he labored diligently to effect a workable agenda for reform. Nothing seemed to escape his attention. Yet, as much as everyday details absorbed his energy, the key to his policy rested on the introduction of fundamental changes in the government's manufacturing program.[19]

Wadsworth always had a special attachment to artillery and, as chief of ordnance, he most often thought and spoke of uniformity in that context. Even before the reorganization of 1815, he had begun to consider means of improving artillery. By 1817 he had proposed the adoption of the British stock-trail system, the effectiveness of which he had witnessed during the War of 1812. To a bureaucracy officially committed to Gribeauval's designs, however, the time was not propitious for such a drastic change. Officers in his own bureau expressed uncertain feelings about the proposal. This uneasiness, coupled with a bitter jurisdictional dispute with the artillerists, forced Wadsworth to satisfy himself with piecemeal efforts to regularize calibers and improve the manufacture of field and siege-garrison equipment under the Gribeauval plan. It was, to be sure, an unsatisfactory arrangement, one that Wadsworth deeply resented. But, given the posture of his peers, he realized that more thoroughgoing reforms would have to wait until petty jealousies and traditions had dissipated. As a consequence, the real thrust of his uniformity policy centered on parallel developments in the manufacture of small arms.[20]

The first indication that the Ordnance Department intended to take action occurred in June 1815, when Wadsworth called a special meeting at New Haven, Connecticut, to discuss the problems of standardizing firearms and to formulate an appropriate strategy. Present were Superintendent Roswell Lee of the Springfield Armory, Superintendent James Stubblefield of the Harpers Ferry Armory, former Springfield superintendent Benjamin Prescott, and Wadsworth's old friend and confidant, Eli Whitney, who hosted the gathering. After several days of deliberation they agreed that uniformity should first be applied to the manufacture of muskets and thereafter extended to all military sidearms. To this end Wadsworth assigned Stubblefield the task of making several pattern pieces at Harpers Ferry based, appropriately, upon the French model musket of 1777. With the completion of this phase, the plan called for further preparations and tests of pattern muskets and rifles and their eventual manufacture as regular products at the national armories. If the experiment proved successful, the program would then be extended to arms made by private contractors.[21]

To his credit, Wadsworth understood that the Ordnance Department could not devise a complex engineering strategy and then simply withdraw as a passive observer of the undertaking. Success depended on closely monitoring and orchestrating every stage of the plan. Since he was overworked and beginning to feel the painful effects of a malignant tumor on his arm, he also knew that he had to select a deputy with the necessary skill and determination to oversee the new program and stand firm in the face of opposition. That person was Lieutenant Colonel George Bomford, an extremely able officer who had graduated from West Point and had served as Wadsworth's principal assistant since 1812. To Bomford fell the responsibility of implementing the uniformity system in America.

Wadsworth and Bomford made strange bedfellows. Although both men came from well-to-do backgrounds and had powerful political connections, no two persons could have been more different in habit and demeanor. Wadsworth was outspoken, Bomford reserved. Wadsworth's mercurial temperament contrasted sharply with Bomford's calculated restraint, just as Bomford's flamboyant lifestyle clashed with Wadsworth's puritan austerity. They frequently disagreed on issues, yet both men realized that, despite their differences, they complemented and strengthened each other. What united them was a zealous commitment to the idea of uniformity and a single-minded determination to see the system introduced on a large scale. Although the cantankerous chief would not live to see the culmination of the mechancial ideal, his ideas would be carried forward and elaborated by his younger protégé. As Wadsworth's designated successor, Bomford would head the Ordnance Department for more than twenty years (1821–42), during which he would witness every important development in the firearms industry. Indeed, from an administrative standpoint, Bomford would do more than any other person to make the uniformity system an American reality.[22]

Lieutenant Colonel George Bomford (1782–1848), chief of the Ordnance Department from 1821 to 1842. (From Records of the Columbia Historical Society of Washington, D.C., *vol. 13 [1910]; Smithsonian Institution Neg. No. 81–193.)*

As architects of the new system, Wadsworth and Bomford pushed relentlessly for the speedy introduction at the national armories. Initially neither officer contemplated serious problems. Their timetable called for completion of the pattern arms by the fall of 1815, with full-scale production of the new "Model 1816" musket to begin at both armories before the turn of the year. When Stubblefield failed to deliver the patterns on time, however, Wadsworth began to realize that neither acceptance nor success would come easily. "It is a Pity that so much Tardiness is manifested in deciding on the Model of the Musket," he wrote Bomford in August 1815. "The Delay looks to me as if the Model I had proposed, or rather the basis which I had laid down for constructing the Model was not altogether approved of." This suspicion was confirmed when the patterns finally reached the Ordnance Department three months later, only to be found deficient in uniformity as well as design.[23]

Sobered by the experience, Wadsworth ordered the preparation of new pattern pieces—this time at both national armories. Moreover, he directed the superintendents to cooperate with one another not only in preparing the patterns but in all matters related to management and manufacturing on the uniformity principle. This instruction, repeated time and again during the next fifteen years, struck a responsive chord with Springfield's Lee, who seized the opportunity to improve interarmory communications by befriending and establishing an ongoing dialog with

Stubblefield, the Harpers Ferry superintendent. Through correspondence and periodic visits both men began to transmit information related to annual inventories, accounting methods, wage rates, and general shop procedures. At Lee's initiative they also exchanged men, machinery, and raw materials, a practice which soon spread to other establishments and had important technological ramifications throughout the industry. Not surprisingly, Lee's resourcefulness, outgoing manner, and disciplined attitude made him one of the most valued members of the American armsmaking community; no one appreciated this more than the chief of ordnance. Under his energetic leadership, Springfield quickly emerged as a leading metalworking center and pivotal clearinghouse for the acquisition and dissemination of technical information. Yet, precisely because he was making the armory "a credit to the Government and an ornament to the nation," he frequently found himself embroiled in affairs that extended far beyond his normal range of activities as a factory master.[24]

Between 1816 and 1819, the Ordnance Department continually reminded Stubblefield and Lee of the need for uniformity in all things and admonished them whenever their efforts fell short of success. Since the craft-oriented armorers of Harpers Ferry resented the intrusion of ordnance officers and often balked at their insistent demands for innovation and change, Stubblefield received the most frequent departmental reprimands. Lee also felt these pressures, however, because Bomford expected him not only to manage Springfield but also to encourage improvements at Harpers Ferry. At times this dual burden became so intense that he threatened to resign his post and seek employment elsewhere. On other occasions he vented his frustration while urging his superiors to be more patient. A typical instance occurred in November 1817, after Bomford had complained for the third time in two years about "the want of uniformity" at the national armories and demanded an immediate remedy. "It is difficult for a Pattern Musket to be made by any one to *please every body,*" Lee responded. "*Faults* will *really exist* and *many immaginary* ones will be *pointed out.* . . . It must consequently take some time to bring about a uniformity of the component parts of the musket at both Establishments." Lee's continued pleas for "time, patience & perserverence" failed to make much of an impression at the Ordnance office. Captivated by the engineering ideal of uniformity and convinced of its urgency, Bomford wanted results, not excuses, and continued to voice criticism whenever the need arose.[25]

Despite these encounters, Lee never wavered in his allegiance to "the grand object of uniformity." Like Bomford and Wadsworth, he understood that the venture's success depended on systematic procedures and regularized controls and that the Ordnance Department served to maintain these standards by closely scrutinizing armory operations. The term "system" meant something very real to these men. Far from being a chimerical symbol, it involved specific regulation of the total production process from the initial distribution of stock to the final accounting of costs. Above all, it embraced the coordination of manufacturing activities between two widely separated arms factories. Like a complex machine— an analogy often used by ordnance officers—the correct functioning of the whole depended on the proper relationship of its parts. Coordination and control thus actuated the systems concept. Much as he disliked the incessant prodding of his military superiors, Lee appreciated the significance of their actions. His realization that uniformity could be achieved only through constant vigilance and scrutiny by a central bureau accentuated the entrepreneurial role of the Ordnance Department and made him a faithful expeditor of its policy.[26]

Ordnance officials relied on two methods of monitoring armory operations. One dealt primarily with fiscal affairs and involved the keeping of accurate accounts; the other addressed problems of quality control and involved the careful inspection of finished firearms. Both methods assumed critical importance in developing and expanding the uniformity system and reflected the influence of uniformity as an organizing concept.

Because every government agency stood accountable to Congress for funds received and expended, the department introduced sophisticated bookkeeping methods at a very early date. Beginning in 1816 all officers in charge of federal arsenals and armories were required by regulations to submit quarterly returns detailing the work performed at their respective posts. Based on standard double-entry bookkeeping, these returns included abstracts and receipts of expenditures for buildings, raw materials, and plant equipment, as well as detailed inventories of work performed and property on hand at the end of each period. The inventories not only identified and tabulated the types of arms manufactured or repaired but also listed components still in progress and the amount of stock—coal, iron, oil, and the like—remaining. In addition, the reporting officer transmitted monthly payroll accounts which recorded the name of each armorer, the type of work he performed, the piece rate for each task, and his total wages. At the end of the fiscal year the superintendent submitted an annual report which summarized the quarterly returns, tallied production, and previewed plans for the coming year. After review by the chief of ordnance and his staff, these records were sent to the Treasury Department for final auditing and approval.[27]

Accurate accounts not only served to justify expenditures and appropriations before a cost-conscious Congress, they also permitted the Ordnance Department to evaluate armory and arsenal operations. By examining the accounts of Springfield and Harpers Ferry in 1820, for instance, Wadsworth discovered that piece rates were much higher at the latter establishment. As a result, he ordered a 12½ percent wage reduction. In like manner, the department used accounts to compare the prices paid for supplies and raw materials at various posts and to alter procurement practices when appropriate. Even more important, accurate bookkeeping provided a means of controlling and coordinating arms inventories throughout the arsenal network, a measure which preceded similar methods in the railroad industry by more than two decades. Precise information on the location, distribution, and condition of equipment among widely scattered arsenals served a strategic purpose. In times of emergency such knowledge enabled officials to direct arms and munitions shipments to the locations where they were most needed; in peacetime it allowed them to replenish depleted stocks and dispose of unserviceable weapons. For these reasons the chief of ordnance paid particular attention to the returns of outlying installations, seeking constantly to improve their accuracy. One benchmark was achieved around 1829 with the complete standardization of all armory and arsenal accounts. Interestingly enough, however, the armories made little effort until the late 1830s to calculate production costs based on allowances for interest, insurance, and depreciation. Even then, ordnance officials used the data only to make rough comparisons of work done at Springfield and Harpers Ferry and not to achieve more effective control over internal production processes. Because they operated in a guaranteed market with output quotas more or less fixed by congressional appropriations, they evidently saw no need to explore further the managerial implications of accurate costing.[28]

Just as accurate bookkeeping enabled the Ordnance Department to survey the general spectrum of production, so regular inspections served to control the exactness of arms being produced. Compared with accounting methods, however, those for inspection evolved slowly. This is not surprising in view of the fact that few ordnance officers originally contemplated the manufacture of fully interchangeable weapons. Wadsworth himself seemed ambivalent about the need for interchangeability when he first formulated the department's policy in 1815. Instead, uniformity was a flexible, open-ended undertaking that developed from a rather crude emphasis on the similarity of parts, to more exacting criteria during the 1820s, to truly interchangeable standards two decades later. Yet even here the picture of a steady evolutionary path toward mechanical perfection is misleading because it obscures retrogressive tendencies that existed in the arms industry throughout the antebellum period. Technological innovation doubtlessly moved forward during these decades but it also was accompanied by temporary setbacks as armorers attempted to digest new mechanical techniques and effectively integrate them into the total production process. Under these circumstances the quality of arms delivered to the government often varied widely even among muskets purportedly of the same model. To remedy these defects the Ordnance Department devised regularized checks against which the work of the national armories and private contractors could be measured. From its endeavors emerged an inspection program which improved quality controls and encouraged the further introduction of mechanized techniques. For this reason inspection methods often paralleled other key technological changes in the industry.[29]

In 1816, the inspection of finished firearms varied not only from one armory to another but also within different branches of the same armory. For the most part the procedure consisted of making qualitative comparisons with a pattern arm and its parts. That is, an inspector discovered work defects mainly by eye rather than by instrument. The basic criteria for inspecting the lock or firing mechanism of a musket, for example, consisted of taking the piece apart and examining its parts to make sure that they had been made in a "workmanlike" manner. If the inspector found the components properly shaped, filed, and finished, he reassembled the piece and tried its action to determine whether it functioned smoothly and gave "good fire." If the lock passed these tests, it received the inspector's stamp of approval. Other than a caliper which served to check the exterior dimensions of the gun barrel and two plugs to verify its bore, no gauges were used during the inspection. The subjective nature of this process, which made it exceedingly difficult to maintain uniformity among muskets of the same model, explains why the chief of ordnance so frequently expressed dissatisfaction with the arms made at the national armories between 1815 and 1821.[30]

The first significant advance in inspection procedures occurred at the Springfield Armory around 1818. There, under the watchful eye of Roswell Lee, master armorer Adonijah Foot and several other workmen developed a method of gauging musket components both during and after the manufacturing process. Although the procedure needed to be perfected, by 1819 it had reached a point of sophistication far in advance of other armories. Within two years Lee could report to Bomford with considerable pride that "our Muskets are now substantially uniform." "Yet," he quickly added, "I am sensible that considerable improvements are yet to be made to complete the system of uniformity throughout all the Establishments."[31]

Long before Lee made this pronouncement, Bomford had been thinking about the larger implications of the Springfield undertaking. Throughout the experiment he had kept in close touch with Lee, continually urging him on. Once the endeavor ended, he applauded the result and in the summer of 1821 announced his intention to introduce Springfield's gauging standard not just at Harpers Ferry but among the private contractors as well. This decision signalled the end of craft-oriented inspection procedures and the beginning of a new mechanical tradition. From that time onward hardened steel gauges would gradually replace human skill in the testing and evaluation of ordnance.[32]

By 1823, the Springfield standard stood in place. In addition to designing an improved pattern musket—sometimes designated the Model 1821—Lee and Stubblefield, acting under Bomford's directive, had prepared and distributed six sets of gauges to various musket contractors. Of the "go-no go" variety, each set consisted of ten different pieces which verified the lock mechanism, the bore and exterior of the barrel, the fall of the stock, the size of the bands, the diameter of the ramrod, and the length and width of the bayonet. To guard against defects produced either by faulty workmanship or wear in the working gauges, Bomford introduced quarterly reinspections of sample weapons produced at both public and private establishments. Different individuals were designated for this duty, although the task most frequently devolved upon the master armorers at Springfield and Harpers Ferry. They carried out the inspections with a master set of gauges, making written reports in which they compared and evaluated the work of the different armories. If any deficiencies existed, they immediately notified the chief of ordnance who, in turn, enjoined those responsible to remedy the situation. At the same time Bomford established a special reference collection of military firearms at Washington for purposes of further comparison and study. He also began to apprise private manufacturers that the issuance of future arms contracts would depend on their current performance, especially the degree to which they updated their operations and cooperated with the department in sharing new inventions or other relevant information. Heeding this injunction, major contractors like Marine T. Wickham, Brooke Evans, and Nathan Starr began almost immediately to adopt new machinery and other laborsaving techniques from the national armories.[33]

Most promising was the work of John H. Hall, a gifted New Englander who had patented a breechloading rifle in 1811, which, with Bomford's support, he had been producing under special contract at Harpers Ferry since 1819. With the aid of over sixty-three inspection gauges and an impressive stable of machinery, Hall conclusively demonstrated in 1826 that his rifles could be made with interchangeable parts, the first of their kind in America. Eight years later Simeon North, an equally talented contractor from Middletown, Connecticut, added another dimension to the evolving pattern of precision when he adopted Hall's gauges and

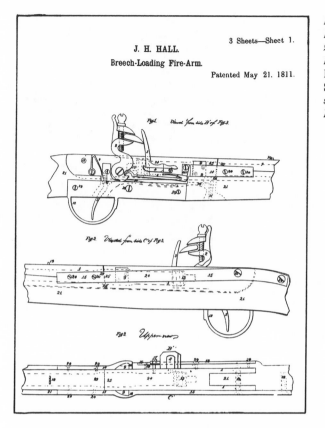

Drawings for John H. Hall's patent of May 21, 1811. (From Claud E. Fuller, The Breech-Loader in the Service *[1933]; Smithsonian Institution Neg. No. 81–195.)*

succeeded in making rifles with parts that could be exchanged with those made by Hall at Harpers Ferry. Together Hall and North provided tangible evidence of what could be accomplished by adopting uniform practices at two widely separated factories. Having personally witnessed the Hall-North operations, Bomford concluded that the uniformity principle had indeed reached a new level of refinement and became even more adamant in his conviction that one day all armories would achieve in a larger, more efficient manner what Hall and North had done on a limited scale. That day would arrive in the mid-1840s when the national armories and private contractors began to produce the Model 1841 percussion rifle and the Model 1842 percussion musket. These were the first fully interchangeable firearms to be made in large numbers anywhere, one of the great technological achievements of the modern era. Yet, as impressive as they were, these accomplishments tended to obscure fundamental ambi-

Simeon North (1765–1852). (Smithsonian
Institution Neg. No. 48698.)

U.S. Rifle Model 1841 (pattern piece)
manufactured at Harpers Ferry Armory.
(From the collections of the Division of
Military History, National Museum of
American History; Smithsonian Institu-
tion Neg. No. 73–312.)

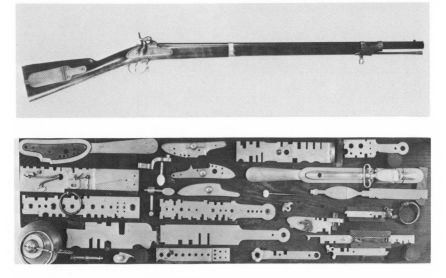

A set of inspection gauges for U.S. Rifle
Model 1841. (From the collections of the
Division of Military History, National
Museum of American History; Smith-
sonian Institution Neg. No. 62468.)

guities and tensions associated with the introduction of the uniformity system.[34]

Throughout the 1820s chronic problems plagued and frustrated the Ordnance Department's policy. For one thing, conflicting opinions existed over the need for mechanization as well as the importance of uniformity. At the main armory at Harpers Ferry, for instance, managers as well as artisans held tenaciously to preindustrial traditions, grumbled about the introduction of new techniques, and vilified Bomford and his "visionary" schemes. Similar feelings also existed among some contractors, although they were less willing to oppose the uniformity policy for fear of losing their contracts. These attitudes bothered Bomford and his staff, but what troubled them more were the lapses that continued to appear in the quality of firearms. Due no doubt to the increasing rigor of inspections, the department received numerous reports of flaws, subterfuges, and shoddy workmanship during the 1820s. By 1827 the problem had become so serious at Harpers Ferry that Lee pleaded with Stubblefield to take "measures... to remedy the deficiency." "The variation between our Arms seems to be greater now than at any time previous," he cautioned. "I hope we shall be able to make some improvement on this point. If not, we shall surely be censured. Let us exert ourselves to the utmost." [35]

No one could deny that Lee had always exerted himself to the utmost, performing services far beyond the normal call of duty. Between 1823 and 1830, special assignments ranging from service on a commission to investigate the establishment of a western armory to twice filling in as acting superintendent at strife-ridden Harpers Ferry had kept him away from Springfield for more than fifteen months. Indeed, in 1823 and again in 1827, he had spent more time away from the armory than at it. But this was not all. In addition to acting as Bomford's chief troubleshooter, Lee also bore responsibility for overseeing the inspection and delivery of all contract arms made in New England. This was a particularly thankless job which involved him in numerous disputes with contractors, hampering his ability to maintain the sort of friendly relations and cooperative contacts necessary for assimilating new techniques at Springfield.

By 1830 these and other activities had sapped Lee's energy and impaired his health. He therefore asked to be relieved of supervision of the contract service. Recognizing that Lee was ill and that he needed to spend more time attending to affairs at Springfield, Bomford acceded to the request by creating the office of Chief Inspector of Contract Arms and appointing a promising lieutenant named Daniel Tyler to the post. The new office placed all small-arms contractors under Tyler's oversight. Tyler performed his job well and the quality of contract arms began to improve. Encouraged by the success, Bomford sought to introduce even more thoroughgoing administrative reforms aimed at consolidating control over the rest of the innovative but erratic industry. Before he could do so, however, an important organizational change had to be made.

In 1821 the Ordnance Department had lost its status as an independent military bureau when Congress, in the interest of economy, had merged it with the Corps of Artillery. The reorganization produced an uneasy union fraught with distrust and ill feeling. Bomford especially chafed under the new arrangement as a "Brevet Colonel of Artillery on Ordnance Service," a title which he considered diminutive. Although the change did not impair developments in the manufacture of small arms, it did retard artillery technology as the two factions jealously guarded their prerogatives and quarreled with one another over the determination of long-range policies. Such rivalry in the field of artillery resulted in stagnation. Young artillerists resented ordnance assignments as an unproductive and unnecessary interruption of their professional careers. Ordnance specialists, on the other hand, became particularly restive about being submerged in an amorphous organization that obfuscated their identity. During the 1820s they repeatedly petitioned the secretary of war to redress the situation. A revision finally came in April 1832, when Congress enacted a bill reestablishing the Ordnance Department as a separate army bureau and returned it to a status comparable to that which had existed in 1815.[37]

The reorganization act of 1832 gave Bomford the authority he needed to effect further managerial changes. As head of an agency charged with administering the procurement, inspection, and distribution of arms over vast geographic distances, he had spent more than a decade addressing problems of a scale and complexity unknown to most contemporaries in the busines world. What is more, he had learned from experience that no individual could personally oversee such widely dispersed yet critically related activities. Since the magnitude of these tasks was so great, the only workable alternative was to develop a decentralized managerial structure through which he could delegate responsibility to subordinates. Daniel Tyler's success as chief inspector of contract arms had demonstrated the effectiveness of this strategy, so in June 1832 Bomford sought to complement that office and expand channels of communication by appointing Lieutenant Colonel George Talcott as inspector of armories, arsenals, and depots. That Bomford chose Talcott for this duty indicates the importance of the office. At the time of his appointment Talcott commanded the Watervliet arsenal near Troy, New York, and was the second ranking officer in the department. In 1848, he would succeed Bomford as chief of ordnance. In fact, from 1832 until the Civil War the inspectorship of armories and arsenals would be filled by senior ordnance officers and all but one would become chiefs of ordnance.[38]

As his title indicated, Talcott's main assignment involved the regular inspection of all federal arsenals and armories, with written reports to his superior, the chief of ordnance. These reports touched on a wide variety of subjects ranging from the condition of buildings and machinery to the general discipline of the establishments. More particularly they compared the operations of different installations, scrutinized their accounts, and evaluated the performance of the superintendents, paymasters, and other supervisory personnel. Being internal documents not intended for public distribution or use, the reports provided remarkably candid appraisals which the chief of ordnance doubtlessly found valuable. In 1832, for instance, Talcott wrote forthrightly of Harpers Ferry that "the machinery at this armory is far behind the state of the manufacture elsewhere and the good quality of their work is effected, at great disadvantage, by manual labor." "Uniformity of work," he emphasized, "is scarcely attainable by manual labor." Three years later he modified his evaluation. "This Armory is much improved since my inspection of 1832—a great advance is perceptible in the introduction of new machinery." "Nevertheless," he quickly added, "much remains to be done to bring it up to the point at which the Springfield Armory stood at that time." Since Springfield was so well managed, it invariably became the model against which the performances of Harpers Ferry and other government arsenals were measured.[39]

The upper waterworks at the Springfield Armory, 1830. (From Claud E. Fuller, Springfield Shoulder Arms, 1795–1865 *[1931]; Smithsonian Institution Neg. No. 48619.)*

Besides reporting on conditions at the institutions they visited, the inspectors of armories and arsenals also recommended various organizational and administrative changes. Moreover, they possessed the authority to make unilateral decisions, as in 1836 when Talcott directed the superintendent at Harpers Ferry to store rejected gun barrels for future use rather than destroy them. Such orders, of course, were subject to review by the chief of ordnance, but he rarely reversed a decision.[40]

Throughout the antebellum period the primary mission of Talcott and his successors centered on the evaluation of production methods and maintenance of uniform standards at all government installations. Indeed, their attempt to impose uniformity in all areas of arms manufacture provided the essential focus which defined the problems and united the efforts of the inspector of armories and arsenals, the chief inspector of contract arms, and even the chief clerk in charge of accounts at the Ordnance Department. By evaluating the work of outlying installations and providing a constant flow of information, these officers relieved the chief of ordnance of a heavy burden. Together they formed a bureaucratic team of middle managers whose separate but interrelated activities vastly improved the internal rhythm of the ordnance enterprise.[41]

The publication in 1834 of an official set of "Ordnance Regulations" formally codified the institutional arrangements that Bomford and Wadsworth had erected during the previous twenty years. In addition to defining the functions of the department, the regulations carefully delineated lines of authority and communications between the chief of ordnance, his representatives, and working personnel in the field. Everyone from prestigious inspectors to the lowliest arsenal custodians thus gained a clear understanding of his respective role and responsibilities within the larger organizational context. At the national armories, for example, the regulations set standards for uniformity of accounting and manufacturing practices and established explicit guidelines for such other matters as employee housing, travel allowances, and work discipline. Similar provisions were established for arsenals, depots, and private contractors. Indeed, nothing seemed to escape the regulations' comprehensive embrace. A decided emphasis on order, control, and accountability distinguished the new rules, which reinforced the Ordnance Department's longstanding commitment to uniformity. By issuing such an elaborate set of instructions and revising them when needed, Bomford and his colleagues formalized an administrative network which greatly facilitated day-to-day operations and clearly identified them with long-range goals. Beyond this, the regulations brought the department public recognition as a central institution of technological change within the government. In doing so, they paved the way for further elaborations of the uniformity system.[42]

A clear indication of the department's growing stature and consolidation appeared in 1839 with the establishment of an investigatory body known as the Ordnance Board. The idea for such a commission was not new. On numerous occasions between 1818 and 1839, the secretary of war had assembled special boards of officers from different branches of the service to investigate and report on model arms, artillery designs, new inventions, administrative changes, and other related matters. What distinguished the new board of 1839 was its convergent membership and its emergence as an ordnance institution. Consisting entirely of ordnance officers, it became the ongoing research and development branch of the Ordnance Department and, within this limited realm, shared important decision-making powers with the bureau chief. Its functions were twofold: "to standardize and systematize Army ordnance equipment." "More than anyone else," one writer observes, the members of the board "were responsible for the development of American military technology during the twenty-two years before the Civil War."[43]

Once established, the Ordnance Board continued the investigations of its predecessors and initiated a number of new ones. The most significant of these undertakings concerned the introduction of a uniform system of artillery. Since 1832, the War Department had made a concerted effort to advance this branch of the service. By 1839 a specific agenda had emerged. Among other things, the secretary of war had sanctioned the introduction of uniform designs and nomenclature for a new field artillery, approved the adoption of stock-trail carriages in place of the outdated Gribeauval plan, and agreed to a complete revision of siege-garrison and seacoast artillery. The introduction of stock-trail carriages presented the fewest problems, since the basic technology for their uniform production already existed in the form of woodworking machinery developed during the 1820s at the national armories. The greatest difficulty centered on finding more uniform methods of making cannon. Interestingly, the solution to this problem involved the Ordnance Board in a protracted series of investigations aimed at determining the "uniform" properties of iron. In this respect, the qualitative concept of uniformity not only informed production practices but also fostered new approaches to problem solving within the department. This, in turn, had another important ramification. Formerly ordnance officers had concerned themselves primarily with promoting and monitoring the uniformity system. By the 1840s they were beginning to become directly involved in the everyday problems of actually building and improving ordnance equipment. Besides denoting a new degree of military involvement in the production process, the endeavor marked the final stage in the rise of the uniformity system.[44]

Derivation of suitable design principles for the new artillery hinged on the production of better gun metal. Since 1801 civilian founders had supplied the army with cast-iron cannon which, for the most part, had served their purposes well. However, so little was known about the physical properties of iron that, whenever cannon burst, no one could adequately explain why. This gap in technical knowlege, coupled with the fact that most European nations refrained from using iron as a construction material, had prompted several military commissions to recommend during the 1830s that the army revert to bronze for light field artillery. Although iron founders who held ordnance contracts firmly opposed the idea and lobbied against it, the secretary of war decided in favor of bronze in 1841 after the Ordnance Board, having recently returned from a nine-month tour of inspection in Europe, persisted in its recommendation. While this action settled the fate of light artillery, a bothersome problem remained. Since bronze was costly and could be used economically only in small artillery pieces, the Ordnance Board had no choice but to continue using iron for heavier siege-garrison and seacoast guns. In reaching this decision, everyone realized that the successful design and construction of large guns depended on producing improved casting which would give safe and durable performance. To attack this problem the board recommended and the War Department approved a special program of industrial research.[45]

The quest for more uniform foundry practices spanned nearly two decades and addressed several separate but related problems associated with the strength of materials. Investigations began in the spring of 1841 when William Wade, a former ordnance officer who had become a proprietor of the Fort Pitt Iron Works in Pittsburgh, returned to government service as a civilian "attending agent" at the private foundries. His position was analogous to the chief inspector of contract arms; his duties, as defined by the Ordnance Board, included the introduction of more uniform production procedures among the contractors as well as the inspection of finished cannon. While executing these tasks, Wade reported to his old friend Bomford who, in 1841, had turned over direction of the Ordnance Department to Colonel Talcott in order to devote full time to the design and improvement of large seacoast artillery known as "Columbiads." During the next ten years and intermittently until his retirement in 1854, Wade spent countless hours conducting comparative trials of cannon, building various gauges and testing machines, and examining fractured samples of iron in an effort to establish correlations between their tensile, torsional, and transverse strength, their specific gravity, and the durability of artillery when subjected to continuous fire.[46]

In carrying out these carefully documented studies, Wade received valuable assistance from several junior officers engaged in similar projects. Most notable was Captain Alfred Mordecai, a West Point graduate and "soldier-technologist" of considerable scientific merit who worked with Wade between 1842 and 1844. Mordecai subsequently achieved prominence for adapting French methods of testing and measuring the explosive force of gunpowder by substituting ballistic and cannon pendulums in place of older eprouvette mortars. His meticulous reports, published in 1845 and 1849, provided incontrovertible evidence that linked the efficiency of gunpowder with its granulation and density. This knowledge, the first of its kind in America, paved the way for further advances in the manufacture of explosives. It also found valuable applications in the planning of fortifications and in the design of cannon, projectiles, and small arms.[47]

The most imaginative series of experiments occurred at the Pikesville arsenal near Baltimore under another protégé of Bomford and Wade, Lieutenant Louis A.B. Walbach. While painstakingly testing and classifying samples of iron from 2,808 cannon during the mid-1840s, Walbach detected "a striking relation between the strength and density of the metal in the same gun" and "the peculiarity of the *color of structure*, and the form and size of the crystals" in the fractured sample. Although he believed these properties could "serve to point out the quality of the iron," certain discrepancies in the observed data prevented him from a definitive statement about their proper relationship. This bottleneck set the stage for an added dimension in ordnance research.[48]

In his final report of 1847, Walbach suggested that chemical analysis might provide an answer to the anomalies he had encountered in testing the physical properties of iron. He therefore sought and eventually received permission to establish a "Chemical Laboratory" at Pikesville. In 1849 he engaged Professor Campbell Morfit of the University of Maryland to study the effects of carbon, sulphur, phosphorus, and other elements on the quality of iron. Working in collaboration with his former teacher, James C. Booth of Philadelphia, Morfit between 1851 and 1855 submitted three reports which adduced metallurgical evidence clarifying and correcting the earlier findings of Wade, Mordecai, and Walbach.[49] Together with the concurrent work of Captain Benjamin Huger, Lieutenant Thomas J. Rodman, and other members of the Ordnance Board, these investigations yielded criteria for evaluating gun metal both before and after the casting process. Although much remained to be learned about the physics and chemistry of iron, the knowledge gained from the experiments enabled founders to exercise far more control over the quality of their castings. Much to the delight of the Ordnance Board, their products significantly improved. Equally important for the growth of modern engineering, the investigations clearly demonstrated the advantages of systematic research over cut-and-try empiricism. By treating the technology of metalworking in this way, members and associates of the Ordnance Department made science a permanent part of the army's research and development program and provided a model for manufacturers in the private sector. In doing so, they stood in the vanguard of those who were beginning to shed craft traditions for a more rigorous approach to the solution of technical problems. This shift of focus and its translation into everyday practice became an essential part of the uniformity system. By the mid-1850s, the impact of the new technology could be seen not only in specialized production techniques like Rodman's sophisticated hollow-casting process but also in the complete standardization of American small arms and artillery.[50]

"My ability," Major Alfred Mordecai confided to his brother in June 1861, "consists in a knowledge and love of order and system, and in the habit of patient labor in perfecting and arranging details; and my usefulness in the Army arises from the long continued application of these qualities to the specialties of my habitual business." Although Mordecai had recently resigned from the service to avoid choosing sides in the fratricidal struggle that enveloped the nation, he could reflect with pride on his many exploits as a "soldier-technologist." For twenty-nine years he had labored diligently as a member of the Ordnance Department. During that time he had served more than two decades on the Ordnance Board; played a leading role in the development of uniform artillery; witnessed the rise of interchangeable manufacturing in the firearms industry; conducted hundreds of experiments with gunpowder, breechloading rifles, time fuses, and other novel weapons; superintended two major arsenals of construction; and administered the affairs of the entire department as a special assistant to the bureau chief. Somewhat ironically, while his distinguished service record had not insured rapid promotion, it had brought other professional rewards. In 1853, he served as judge of exhibits at New York's Crystal Palace exposition. The same year he was elected to membership in the prestigious American Philosophical Society and invited to join the American Association for the Advancement of Science. To be sure, these honors acknowledged Mordecai's many individual achievements. In a larger sense they also recognized the institutional accomplishments of the Ordnance Department and at both levels symbolized the growing link between military innovation and the civilian world of science and industry.[51]

Few colleagues could match Mordecai's breadth of experience and impressive record of achievement. Yet his "love of order and system" and "habit of patient labor" reflected a pervasive spirit which had inspired and guided the ordnance enterprise since its inception. Order and system, patience and perseverance, pragmatism and perfectability, simplicity and solidarity: these value-laden precepts formed a common intellectual bond between members of the department and gave them a keen sense of professional identity. From this ethos had emerged the commitment to pursue a policy of uniformity as well as the need to make it a collective managerial endeavor.

Scholarly treatments of entrepreneurship are usually confined to individuals or groups operating within the private business arena. This focus is understandable in view of the importance that is assigned to risk taking. Yet even though antebellum ordnance officials cannot be labeled entrepreneurs in the strictest sense of the word, they nonetheless exercised important entrepreneurial functions in managing the armory-arsenal complex, in making critical decisions, and—not to be taken lightly—in lobbying and accounting for congressional appropriations. Introduction of the uniformity system provided the context for these activities, involving officers like Mordecai in a complex innovative process that moved through at least six interactive stages of problem identification, idea response, invention, research and development, introduction into use, and diffusion to other users. Each stage formed a particular pattern that impressed itself on the structure of the whole, but the one which best reflected the department's peculiar style of innovation was the process of diffusion.[52]

As early as 1815 the chief of ordnance and his staff had recognized the importance of diffusing new techniques. Without the rapid assimilation of machine and gauging processes, the coordination and control necessary for uniform production would have proved well-nigh impossible. Continual communication and cooperation therefore became essential elements in the department's strategy. Resultant information transfers soon manifested themselves in published reports of tests and experiments and in the day-to-day operations of the armaments network.

Nowhere was the emphasis on communication and cooperation stronger than in the small-arms industry. There, as indicated earlier, private contractors as well as federal administrators regularly corresponded, visited, and assisted one another in working out the basic configurations of the uniformity system. To insure that the diffusion process would not be hindered, the Ordnance Department also insisted that the national armories open their shops to all visitors who could make drawings, borrow patterns, and obtain other information pertinent to their special interests. At the same time, the department had an implicit understanding with all arms contractors that they had to share their inventions with the national armories on a royalty-free basis if they wished to continue in government service. In this way a number of novel metalworking and woodworking techniques that had originated in the private armories of persons like

Simeon North became part of the public domain. Such an "open door" policy, while accentuating the public-service orientation of the Ordnance Department, explains why so few crucial machines and machine processes were actually patented during the antebellum period. The same policy also accounts for the highly integrated character of the American firearms industry as well as for the relative speed with which manufacturers adopted the new technology.

Although ordnance officials well understood the importance of diffusing mechanized techniques within the arms industry, they did not anticipate the far-reaching consequences the process would have for other sectors of the economy. By the mid-1840s the manufacturing methods aggregated at the national armories had begun to spread beyond new armories like the Robbins & Lawrence works in Windsor, Vermont, to all kinds of factories and machine shops. To a certain extent this change manifested itself through former ordnance officers who had assumed managerial posts with various firearms, foundry, and railroad businesses. Far more frequently, however, armory practice was transmitted by workmen who had received their early training at one of the public or private armories and subsequently moved to new positions as master machinists and production supervisors at other manufacturing establishments. Typical in this respect was Jacob Corey MacFarland, a skilled machinist who left the Springfield Armory around 1845 to become the foreman of the Ames Manufacturing Company's machine shop at Chicopee, Massachusetts.

The experience of the Ames company well illustrates how firms availed themselves of the new technology. Established in 1834 by the brothers Nathan P. and James T. Ames, the company became one of the earliest in the United States to manufacture a standard line of machine tools. As a mixed enterprise operating in a limited market, the firm undertook special jobbing contracts and made a wide variety of millwork, mining equipment, cutlery, small arms, cannon, statuary, and other metal products. Yet despite their obvious technical skill and versatility, the Ames brothers were basically copyists rather than innovators. And for good reason. Chicopee was only a few miles from Springfield, where the brothers had ready access to patterns and drawings owned by the national armory. Aided by a number of former Springfield workmen, their stock of machinery clearly reflected this influence.[53]

Other transitional establishments—Colt, Robbins & Lawrence, the Sharps Rifle Company, E. Remington & Sons, George S. Lincoln & Company, and the Providence Tool Company, to name but a few—followed similar practices, thereby acquiring the latest armory know-how and relieving themselves of lengthy time lags and costly expenditures associated with learning-by-doing. Frequently borrowed tool and machine designs had to be adapted to different uses. This in turn led to further improvements and inventions. Coupled with the aggregate growth of factory production, such activities serve to explain why the late antebellum period witnessed so many significant advances in machine design and production.

From this complex tapestry of cooperation, diffusion, and innovation, one can discern a mechanical genealogy that directly links the Springfield Armory and the New England arms industry with the rise of the modern machine-tool industry. Together these institutions and their cadres of remarkably mobile mechanics fostered a quasi-educational phenomenon one writer aptly describes as "technological convergence." Through them, armory practice spread to technically related industries and by the late 1850s could be found in factories making sewing machines, pocket watches, railroad equipment, wagons, and hand tools. From these beginnings it was only a matter of time before the new technology found applications in the production of typewriters, agricultural implements, bicycles, gramophones, cameras, automobiles, and a host of products associated with the mass-production industries of the twentieth century. Interestingly, machining processes rather than precision instrumentation constituted the most important transfer from the arms industry. Since precision production was expensive, most businessmen contented themselves with manufacturing highly uniform but not necessarily interchangeable parts. Only the government could afford the luxury of complete interchangeability.[54]

The naive enthusiasm with which many midnineteenth-century writers celebrated the advent of the machine tends to obscure the inherent difficulties that accompanied the early industrialization of America. Mechanization, despite its many practical benefits, exacted a costly social and psychological toll. The very process of factory innovation uprooted a large segment of the nation's population, catapulting it into a vastly different, bureaucratically organized environment. None felt the impact of these changes more forcefully than the people who lived and worked in the mill communities. Yet, as much as churchlike factories and whirling machines defined and symbolized the new order of things, what really distinguished the age were the unsettling changes effected by the division of labor. In this respect, the experience of the national armories was no different from other large industrial establishments in the United States.

When labor was divided in nineteenth-century arms factories, individual work assignments became more simplified while the overall production process became more complex. Coordinating and controlling the flow of work from one manufacturing stage to another therefore became vital and this, in turn, demanded closely regulated on-the-job behavior. Under these conditions the engineering of people assumed an importance equal to the engineering of materials. As conformity supplanted individuality in the workplace, craft skills and other preindustrial traditions became a detriment to production. To those who planned and orchestrated the uniformity system, such changes—though initially unanticipated—seemed fully compatible with their ideas of rational design and with their paternalistic concepts of Christian stewardship. To those who worked under the system, the new regimen represented a frontal assault on valued rights and privileges. The resulting confrontations between labor and management over the control of work highlight fundamental tensions that seethed beneath the glowing veneer of industrial achievement. Therein lay the subtle nuances of technology as a social process.[55]

Just as the general regulations of the Ordnance Department sought to coordinate and control the federal network of armories and arsenals, so formal work rules aimed at governing the daily activities of their employees. The process began in 1816 when Roswell Lee, at the behest of the chief of ordnance, issued a set of twenty-three regulations for the Springfield Armory. The very nature of these rules suggests the sorts of problems Lee and his predecessors had experienced in getting workers to abandon customary habits and reorder their lives around values of diligence, thrift, and responsibility. Among other things, the regulations forbade scuffling, playing, fighting, gambling, drawing lotteries, drinking "ardent spirits," and making "any indecent or unnecessary noise" during working hours. Besides enjoining the armorers "not to begin, excite or join in any Mutinous, riotous or Seditious conduct," the rules also stipulated that "due attention is paid to the Sabbath and no Labor, Business, amusement, play, recreation... or any proceeding incompatible with the Sacred Duties of the day will be allowed." Those who transgressed these norms were chastized and, in some cases, fined. Repeat offenders were fired and blacklisted, their chances of finding employment elsewhere in the region virtually eliminated. Not surprisingly, these and other sanctions established a pattern of paternalism at Springfield that closely paralleled similar practices in the New England textile industry. And, as in the textile mills, enforcement of the ordnance regulations proved difficult.[56]

Lee's success in making Springfield a showplace of American industry owed much to his ability to impose regulations without alienating his labor force. With his death in 1833, however, the situation changed. Discipline relaxed under his successor, John Robb. By 1841 a special commission appointed to examine "the condition and management" of the armory noted many irregularities in the daily operations of the establishment. In their published report the members expressed "surprise... that no regular hours were established for labor." "Every mechanic, working by the piece," they noted, "is permitted to go to his work any hour he chooses, and to leave off at his pleasure."

In some instances the machinery at the water-shops has been kept running for the accommodation of a single mechanic; and in most of the visits of the board, though made in hours usually devoted to labor, these shops were found nearly deserted. The reading of newspapers during the ordinary hours of labor appears to be so common a practice as not to be deemed improper.

Indeed, the report bristled, "the reading was continued even during the inspection of the board." That many armorers could labor only a few hours daily and still earn relatively high wages of $40 to $60 a month convinced the commissioners that "there has been great looseness in the management of the armory; and although the machinery has been brought to a high degree of perfection, and the work done is generally creditable to the mechanics, yet these results have been attained by a lavish expenditure of the public money." Such conditions, they concluded, "could not exist in a private [business], and ought not to be permitted in a public establishment." [57]

The irregularities detected at Springfield seemed to fade in significance when compared with affairs at Harpers Ferry. For years the Potomac armory had suffered the reputation of being locally controlled, shamefully abused, and flagrantly mismanaged. It also held the dubious distinction of employing one of the most independent and troublesome labor forces in the country. Attempts to introduce work rules during the 1820s generally went unheeded. "Workmen came and went at any hour they pleased," one officer recalled, "the machinery being in operation whether there were 50 or 10 at work." Along with these practices, armorers claimed the privileges of keeping frequent holidays, transferring jobs at will, drinking whiskey on the premises, and selling their tools "as sort of a fee simple inheritance." They also boasted that anyone who interfered with these rights could expect the same fate that befell Thomas Dunn, the one superintendent who had adopted and rigorously enforced Lee's regulations. In 1830, after only six months in office, Dunn had been murdered by a disgruntled armorer. Everyone knew that the workers determined their own standards of conduct and that they did so with the connivance and support of community leaders and local politicians. "Every way considered," a newly appointed master armorer wrote in amazement to a friend, "there are customs and habits so interwoven with the very fibers of things as in some respects to be almost hopelessly remitless." [58]

Ironically, the very innovations that made the uniformity system possible contributed to the perceived "labor problem" at Springfield and Harpers Ferry. The introduction of self-acting machinery enabled armorers paid by the piece to complete more work in less time. Rather than trying to maximize their incomes by making optimal use of the new technology, however, most of them chose to limit their output to customary levels and carry home approximately the same monthly wages. Such practices, unhampered by managerial constraints, gave them the leisure to pursue other occupations such as farming and allowed them to maintain the relatively high level of piece rates previously paid for hand labor. Thus, instead of lending itself to the rationalization of production, unregulated piece work became the means by which armorers controlled their working hours and retained other preindustrial traditions in the midst of rapid technological change.

Committed as they were to "stability of things and stability of mind" in all undertakings, members of the Ordnance Department deplored the instability that characterized labor practices at the national armories. As ranking officers in the department, Bomford and Talcott felt especially embarrassed and frustrated about the situation. On numerous occasions during the 1830s they had cautioned superintendents about the lack of internal discipline and had urged them to institute reforms. Beyond this, however, they could do little to enforce the regulations because they resided so far away from the armories and because the highest officials at Harpers Ferry and Springfield—superintendents, paymasters, and master armorers—were civilians who held their appointments through the machinations of the patronage process. Accordingly, political interests and considerations informed their actions as managers. Since their local power and support depended on maintaining good will in their respective communities, they rarely flaunted their authority or did anything that threatened to jeopardize their relations with the armorers. Like good politicians, they catered to the interests of their constituents and, with regard to work regulations, left well enough alone.[59]

Thwarted in many attempts to control the civilian superintendents, ordnance officials had long been looking for an opportunity to initiate sweeping administrative changes at the national armories. That time arrived in 1841 with the inauguration of President William Henry Harrison. When the newly appointed secretary of war, John H. Bell, announced his intention to introduce a thorough agenda of reform and called upon the chief of ordnance for advice, Bomford recommended replacing civilian superintendents with ordnance officers at the armories. Bell agreed to the plan and Bomford immediately ordered two of his most experienced subordinates to take command at Springfield and Harpers Ferry. Although patronage-conscious politicians felt threatened by the arrangement and roundly denounced it as "full of mischief in *all respects*," the secretary nonetheless remained firm in his commitment and the department's "new reign" began on April 16, 1841.[60]

During the next thirteen years military superintendents exercised exclusive control over the internal operations of both national armories. At their insistence workers abandoned the task-oriented world of the craft ethos and reluctantly entered the time-oriented world of industrial capitalism. That the large-scale manufacture of interchangeable firearms paralleled this change was no mere coincidence. From the outset ordnance officers recognized the importance of work rules, clocked days, and regularized procedures in stabilizing the complex human and physical variables present in the workplace. Experience had taught them that there was no other alternative: a factory discipline characterized by rigid bureaucratic constraints had to be inculcated and absorbed by all employees. Only in this way could the delicate parameters of the uniformity system be maintained.

Such considerations, of course, gave little solace to the many armorers whose work ways were being altered. Time and again pieceworkers and inspectors complained about having to keep regular hours. Time and again they grumbled about the relentless pressures imposed by the new administration as well as the rigor with which it enforced the rules. While older artisans bemoaned the disappearance of traditional skills, other armorers protested against the installation of timeclocks and the lowering of piece rates. All these feelings were reinforced by kinsmen and neighbors who customarily distrusted strangers and resented outside meddling in their local affairs. Since the culprits were military men, politicians from both regions continually fanned discontent by publicly attacking the "despotism and oppression" of military rule.[61]

Not surprisingly, members of the Ordnance Department denied these charges and repeatedly asserted that the change from civilian to military superintendents "produced a great, if not entire reformation of the abuses formerly existing" at the armories. As Bomford's successor, Talcott became deeply involved in the dispute and often found himself clarifying and defending the department's position. "The regulations governing the workmen are *not changed*," he wrote the secretary of war on one occasion, "they are *merely enforced*." Indeed, he declared, "the *real ground of opposition* to the present mode of supervision is well known to be this":

> The men have been paid high prices & were in the habit of working from 4 to 6 hours per day—& being absent whole days or a week. At the end of a month their pay was generally the same in amount as if no absence had occurred. They are now required to work full time and during fixed hours (according to old regulations) and the master of the Shop keeps

a time account showing the time *actually spent in labor*. Here is the *great oppression* complained of. At the end of a month the *quantity of labor performed*, or *product*, and the *time* during which it is effected, are seen by a simple inspection of the Shop books. The degree of diligence used by each man is also known and hence results a knowledge of what is the *fair price* to be paid for piece work!!! The Armorers may attempt to disguise or hide the truth under a thousand clamors—but this is the *real cause* of their objections to a Military Superintendent. He enforces the Regulations which lay bare their secret practices (frauds—for I can use no better term)....[62]

Although Talcott's spirited rejoinders delineated the sources of antagonism and ill-feeling at the armories, there was no denying the close association that existed between military discipline and work discipline. What pained him most was the charge of "despotism and oppression," terms which harkened back to the Revolutionary era and still conveyed deep meaning in Jacksonian America. Such epithets portrayed the Ordnance Department as a threat to the enduring ideals of republicanism. Yet, as Talcott and his colleagues constantly asserted, nothing could have been further from the truth. Like other industrialists, they thoroughly agreed with an 1846 pronouncement by New York's Matteawan Company that systematic control of the work process was necessary "to convince the enemies of domestic manufactures that such establishments are not 'sinks of vice and immorality,' but, on the contrary, nurseries of morality, industry, and intelligence." Beyond this, they firmly believed that an orderly and well-regulated work environment would not only promote efficiency but also instill values conducive to the moral growth and well-being of the country. Viewed in this context, the enforcement of armory discipline had a much larger cultural purpose.[63]

The identification of godliness with productivity melded perfectly with the guiding principles of the Ordnance Department. By amalgamating the sacred and the profane, members of the department sought to clothe utilitarianism in the garb of republican simplicity and thus legitimize their efforts. In their minds the uniformity system meant something more than a means of production; it represented a barometer of national development, an achievement in which Americans could take pride and from which republican institutions could draw strength and vitality. In this sense, uniformity not only stood for material progress through technology but also for the maintenance of virtue in an era of developing industrialism.

It is impossible to specify exactly the economic costs of introducing the uniformity system. Since the new technology evolved in increments closely associated with daily manufacturing activities, existing records simply do not distinguish between developmental and production costs. An admittedly rough approximation of expenditures for special tools, equipment, raw materials, labor, administration, and shop space as well as interest, insurance, and other hidden capital costs, suggests that the development of uniform standards required a total investment exceeding two million dollars. Although private contractors expended time and money in developing new techniques, only the federal government could have financed such a massive undertaking. That it did so over a forty-year period underscores the importance not only of capital but also longevity as a key ingredient in the evolution of complex technological systems. What the government provided, in addition to large infusions of money, was an ongoing bureaucratic organization within which the new technology—itself a bureaucratic phenomenon—could evolve. That the innovation transcended both individual limitations and the confines of isolated geographic environments was due largely to agency of the Ordnance Department.

Notes

1. For entry to the literature on this subject, see Robert A. Lively, "The American System: A Review Article," *Business History Review* 29 (1955):81–96; Sidney Fine, *Laissez Faire and the General Welfare State* (Ann Arbor, Mich., 1956), chap. 1; Henry W. Broude, "The Role of the State in American Economic Development, 1820–1890," in *The State and Economic Growth*, ed. Hugh G.J. Aitken (New York, 1959), pp. 4–22; David D. Van Tassel and Michael G. Hall, eds., *Science and Society in the United States* (Homewood, Ill., 1966), chaps. 8–9; Carter Goodrich, "Internal Improvements Reconsidered," *Journal of Economic History* 30 (1970):289–311; and Harry N. Scheiber, "Government and the Economy: Studies of the 'Commonwealth' Policy in Nineteenth-Century America," *Journal of Interdisciplinary History* 3 (1972):135–54.

2. For the undertakings of the Topographical Corps, see William H. Goetzmann's prize-winning study of *Army Exploration in the American West* (New Haven, Conn., 1959). For the Corps of Engineers, see Forest G. Hill, "Government Engineering Aid to Railroads Before the Civil War," *Journal of Economic History* 11 (1951):235–46; Hill, *Roads, Rails and Waterways: The Army Engineers and Early Transportation* (Norman, Okla., 1957); and Hill, "Formative Relations of American Enterprise, Government and Science," *Political Science Quarterly* 75 (1960):400–19.

3. The original report appeared in the British *Parliamentary Papers* 50 (1854–55) and is reprinted in *The American System of Manufactures*, ed. Nathan Rosenberg (Edinburgh, 1969), p. 143. James Nasmyth also used the

expression in his *Autobiography* (1883) with specific reference to developments at the Springfield Armory. Joseph W. Roe later quoted Nasmyth and adopted the term in his pathbreaking history of *English and American Tool Builders* (New Haven, Conn., 1916), pp. 129, 140–41. One of the earliest American writers to refer to the "American system" was Charles H. Fitch in the section on "The Manufacture of Fire-Arms," in "Report on the Manufactures of Interchangeable Mechanism," in *Report on the Manufactures of the United States at the Tenth Census* (Washington, D.C., 1883).

4. See Letters Sent-Letters Received and Reports of Inspections of Arsenals and Depots in the Records of the Office of the Chief of Ordnance, Record Group 156, National Archives, Washington, D.C. (hereafter cited as OCO); and Stephen V. Benet, ed., *A Collection of Annual Reports and Other Important Papers, Relating to the Ordnance Department*, 3 vols. (Washington, D.C., 1878–90).

5. Michael Kammen, "From Liberty to Prosperity: Reflections Upon the Role of Revolutionary Iconography in National Tradition," *Proceedings of the American Antiquarian Society* 86 (1976):263–72; Russel B. Nye, *Society and Culture in America, 1830–1860* (New York, 1974), pp. 1–31; Leonard G. White, *The Jeffersonians: A Study in Administrative History, 1801–1829* (New York, 1965), pp. 211–39.

6. U.S. *Statutes at Large*, 2:732–34; White, *The Jeffersonians* (note 5), pp. 215–16, 224–32. For an analysis of small-arms procurement before 1812, see Merritt Roe Smith, "Military Arsenals and Industry Before World War I," in *War, Business, and American Society: Historical Perspectives on the Military-Industrial Complex*, ed. B. Franklin Cooling (Port Washington, N.Y., 1977), pp. 25–28.

7. U.S. *Statutes at Large* 3:203–5. Stanley L. Falk describes these changes as a "Technological Revolution in military arms and equipment" in "Soldier-Technologist: Major Alfred Mordecai and the Beginnings of Science in the United States Army" (Ph.D. diss., Georgetown University, 1959), p. 1 et passim.

8. See, for example, Eugene S. Ferguson, "On the Origin and Development of American Mechanical 'Know-How'," *Midcontinent American Studies Journal* 3 (1962):3–15; Norman B. Wilkinson, "Brandywine Borrowings from European Technology," *Technology and Culture* 4 (1963):1–13; Paul J. Uselding, "Henry Burden and the Question of Anglo-American Technological Transfer in the 19th-century," *Journal of Economic History* 30 (1970):312–37; and David J. Jeremy, "British Textile Technology Transmission to the United States," *Business History Review* 47 (1973):24–52.

9. See, for example, Ralph R. Shaw, *Engineering Books Available in America Prior to 1830* (New York, 1933); Edward C. Ezell, "The Development of Artillery for the United States Land Service Before 1861: With Emphasis on the Rodman Gun" (Master's thesis, University of Delaware, 1963), pp. 80–82, 122; Peter M. Molloy, "Technical Education and the Young Republic: West Point as America's École Polytechnique, 1802–1833" (Ph.D. diss., Brown University, 1975); Benet, *Ordnance Reports* (note 4), 1: 18–19, 29, 53–57, 185, 202, 214, 224, 233–34, 510. For references to the "French system

of artillery," see William E. Birkhimer, *Historical Sketch of the Organization, Administration, Materiel and Tactics of the Artillery, United States Army* (1884; reprint ed., New York, 1968), pp. 225, 235; and Louis de Tousard, *American Artillerist's Companion*, 3 vols. (Philadelphia, 1809), 1:200.

10. *Dictionary of American Biography*, s.v. "Tousard, Anne Louis de"; Norman B. Wilkinson, "The Forgotten 'Founder' of West Point," *Military Affairs* 24 (1960–61):177–8; Ezell, "Development of Artillery" (note 9), pp. 79–87.

11. Tousard, *Artillerist's Companion* (note 9), 1:v–vi; 2:iii–iv, xiv; Birkhimer, *Historical Sketch of Artillery* (note 9), p. 302; William Wade, "Early Systems of Artillery," in *Ordnance Notes, No. 25* (Washington, D.C., 1874), pp. 140–41. Another influential French tract was Colonel Jonathan Williams's translation of De Sheel's *A Treatise on Artillery* (1800).

12. Captain Ralph Willet Adye, quoted by Tousard, *Artillerist's Companion* (note 9), 1:200; 2:xiii–xvi, xviii. "In order to attain this desired uniformity," Tousard observed, "a certain number of patterns for the length and figure of the guns; for the dimensions and proportions of the carriages and appendages; and for the best manner of training them, was irrevocably fixed by regulations. Exact tables were formed, and accurate scales of the dimensions of every article used in artillery, made and marked with the seal of the government. Models for workmen in wood; patterns, matrices, and dies, for the forges, were made in conformity to those measures to insure the regularity of the principal forms in the carriage work. These were placed in every arsenal, and *officers made accountable for their construction and reception.* Then the expense eventually lessened, evinced the advantage of having adopted a system, and of having introduced economy, precision, and uniformity in this service." It is noteworthy that American ordnance officers used virtually the same language and argument in explaining and defending their plans for uniformity before congressional committees during the early nineteenth century.

Studies assessing the origins and technological implications of the Gribeauval and British stock-trail systems are needed. Selma Thomas is currently investigating the origins of the French uniformity system. For a preview of her work, see " 'The Greatest Economy and the Most Exact Precision': The Work of Honoré Blanc" (paper delivered at the Twentieth Annual Meeting of the Society for the History of Technology, Washington, D.C., 21 Oct. 1977). I am endebted to Ms. Thomas for providing a copy of her paper and sharing her research findings on Gribeauval.

13. Birkhimer, *Historical Sketch of Artillery* (note 9), pp. 227–49.

14. Blanc, *Mémoire Important sur la fabrication des armes de guerre, à l'Assembleé Nationale* (Paris, 1790), quoted by Thomas (note 12), p. 1. For discussions of Blanc's techniques, see William F. Durfee, "The First Systematic Attempt at Interchangeability in Firearms," *Cassier's Magazine* 5 (1893–94):469–77; Roe, *English and American Tool Builders* (note 3), pp. 129–30; Edwin A. Battison, "Eli Whitney and the Milling Machine," *Smithsonian Journal of History* 1 (Summer 1966):11–26; and Merritt Roe Smith, *Harpers Ferry Armory and the New Technology* (Ithaca, N.Y., 1977), pp. 88–89, 91, 232.

15. Jeanette Mirsky and Allan Nevins, *The World of Eli Whitney* (New York, 1952), pp. 208, 214, 219–22; Robert S. Woodbury, "The Legend of Eli Whitney and Interchangeable Parts," *Technology and Culture* 1 (1960):243–44; Edwin A. Battison, *Muskets to Mass Production* (Windsor, Vt., 1976), pp. 8–9. For a recent evaluation of Whitney's accomplishments, see Merritt Roe Smith, "Eli Whitney and the American System of Manufacturing," in *Technology in America: A History of Individuals and Ideas*, ed. Carroll W. Pursell (Cambridge, Mass., 1981).

16. Mirsky and Nevins, *Eli Whitney* (note 15), pp. 195–96, 200, 206–7, 261, 267–68, 291; Benet, *Ordnance Reports* (note 4), 1:15–17; C. Wingate Reed, "Decius Wadsworth, First Chief of Ordnance, U.S. Army, 1812–1821," *Army Ordnance* 24 (May-June 1943):527–30, Part 1; 25 (July-Aug. 1943):113–16, Part 2; Birkhimer, *Historical Sketch of Artillery* (note 9), p. 278; Wadsworth to secretary of war, 8 Aug. 1812, Letters Sent to the Secretary of War, OCO; Whitney to Roswell Lee, 10 Nov. 1821, Letters Received, Records of the Springfield Armory, Record Group 156, National Archives (hereafter cited as SAR).

17. Wadsworth, 27 May 1813, quoted by Birkhimer, *Historical Sketch of Artillery* (note 9), p. 386.

18. Ibid., pp. 277–78; Benet, *Ordnance Reports* (note 4), 1:5–6; U.S. *Statutes at Large*, 3:203–5.

19. See, for example, Benet, *Ordnance Reports* (note 4), 1:19–79. By 1821, the year of Wadsworth's death, the Ordnance Department maintained three "arsenals of construction at Watervliet, New York; Pittsburgh, Pennsylvania; and Washington, D.C. Of the twenty-three arsenals reporting to the department in 1855, four (Watervliet, Allegheny, Washington, and St. Louis) were "arsenals of construction," six were "arsenals of repair," and thirteen were "arsenals of deposit."

20. Ibid., 1:29, 34–36, 53–55, 58–61, 71–72; Reed, "Decius Wadsworth" (note 16), Part 1, p. 530, Part 2, p. 115; Birkhimer, *Historical Sketch of Artillery* (note 9), pp. 230–37, 260–62; Ezell, "Development of Artillery" (note 9), pp. 95–100; Falk, "Soldier-Technologist" (note 7), pp. 231–33; Wade, "Early Systems of Artillery" (note 11), pp. 141–42, 148.

21. Wadsworth to Colonel George Bomford, 15 May and 13 June 1815, Stubblefield and Lee to Wadsworth, 30 Mar. 1816, Correspondence Relating to Inventions, OCO; Lee to Stubblefield, 6 Aug. 1816, Letters Sent, SAR.

22. For further information on Bomford (1780–1848), see Smith, *Harpers Ferry Armory* (note 14), p. 153.

23. Wadsworth to Bomford, Aug. 1815, Correspondence Relating to Inventions, OCO.

24. Stubblefield and Lee to Wadsworth, 30 Mar. 1816, Correspondence Relating to Inventions, OCO; Lee to Stubblefield, 6 Aug. 1816, 10 and 20 Nov. and 18 Dec. 1817; 18 June and 11 July 1818, Letters Sent, SAR; *American State Papers: Military Affairs*, 2:553.

25. Lee to Stubblefield, 20 Nov. 1817, to Bomford, 11 Sept. 1821, Letters Sent,

SAR; Lee to Senior Officer of Ordnance, 20 Nov. 1817, Letters Received, OCO.

26. *American State Papers: Military Affairs*, 2:543–44, 552–53; Lee to Asa Waters, 8 June 1819, Letters Sent, SAR.

27. See payrolls and accounts of U.S. armories and arsenals, 1816–50, Second Auditor's Accounts, Records of the United States General Accounting Office, Record Group 217, National Archives.

28. Smith, *Harpers Ferry Armory* (note 14), p. 152; Felicia J. Deyrup, *Arms Makers of the Connecticut Valley: A Regional Study of the Economic Development of the Small Arms Industry, 1798–1870*, Smith College Studies in History 33 (Northampton, Mass., 1948), pp. 49–51, 131; Bomford to Lee, 25 July 1829, Letters Sent, OCO; Lee to Thomas B. Dunn, 16 Jan. 1830, Letters Sent, SAR. On the managerial innovations of the Springfield Armory and trunkline railroads, see Alfred D. Chandler, Jr., *The Visible Hand: The Managerial Revolution in American Business* (Cambridge, Mass., 1977), pp. 72–75, 79–121.

29. Gene S. Cesari, "American Arms-Making Machine Tool Development, 1789–1855" (Ph.D. diss., University of Pennsylvania, 1970), and Deyrup, *Arms Makers* (note 28) document these changes. Also see Fitch, "The Manufacture of Fire-Arms" (note 3), p. 4.

30. Stubblefield to Bomford, 9 Nov. 1821, Letters Received, OCO.

31. *American State Papers: Military Affairs*, 2:544; Lee to Bomford, 11 Sept. 1821, Letters Received, OCO.

32. Lee to Bomford, 11 Sept. 1821, OCO; Lee to Stubblefield, 29 Sept. 1821, Letters Sent, SAR; Stubblefield to Lee, 12 Oct. 1821, Letters Received, SAR.

33. Stubblefield to Bomford, 9 Nov. 1821, Stubblefield and Lee to Bomford, 4 Dec. 1821, Letters Received, OCO; Lee to Stubblefield, 15 Dec. 1821, to Bomford, 18 Dec. 1821, 23 Nov. 1822, to Eli Whitney, 13 Nov. 1822, Letters Sent, SAR; Stubblefield to Lee, 17 Nov. 1821, 2 Aug. 1822, Letters Received, SAR; Bomford to Stubblefield, 15 Aug. 1827, Letters Sent, OCO; Deyrup, *Arms Makers* (note 28), pp. 56–59. Also see Asa Waters to Bomford, 6 Apr. 1824, Letters Received, OCO; Bomford to Nathan Starr, 19 Nov. 1827, Letters Sent, OCO; Starr to John H. Eaton, 10 July 1829, Correspondence Relating to Inventions, OCO; James Baker to Lee, 6 Apr. 1826, Marine T. Wickham to Lee, 18 Feb. 1829, Letters Received, SAR.

34. These developments are fully treated in Smith, *Harpers Ferry Armory* (note 14), pp. 184–251, 280–92. Increasingly sophisticated gauging standards continued to parallel mechanized manufacturing developments throughout the antebellum period. During the early 1840s, for instance, sixty-eight gauges—many of them multiple gauges—were used in the inspection of contract arms. By 1854, their number had increased to nearly 100 and during the Civil War the Springfield Armory would use over 150 gauges in testing the accuracy of component parts. See Deyrup, *Arms Makers* (note 28), p. 146; and Rosenberg, *The American System* (note 3), pp. 143, 187–90. The accuracy of these methods received an unexpected test in 1852 when, as a result of a flood at

Harpers Ferry, 9,000 percussion muskets with unmarked parts were stripped, cleaned, and randomly reassembled "with every limb filling and fitting its appropriate place with perfect exactness." For further information, see Benet, *Ordnance Reports* (note 4), 2:536.

35. Lee to Stubblefield, 31 Oct. 1827, Letters Sent, SAR. For a sampling of inspection reports on contract as well as national armory firearms, see inspection report of Philip Hoffman, 24 Feb. 1823, Adonijah Foot to Bomford, 10 June 1823, extracts from the confidential report of Colonel Archer, 1824, Elisha Tobey to Lee, 27 Oct. 1827, Letters Received, OCO; Lee to Bomford, 6 June 1826, Joseph Weatherhead to Bomford, 22 May 1827, Letters Sent, SAR.

36. *American State Papers: Military Affairs*, 2:731; Smith, *Harpers Ferry Armory* (note 14), pp. 171–74, 178–81; William Riddal to Lee, 21 Dec. 1829, Letters Received, SAR; Lee to James Carrington, 4 Aug. 1830, Letters Sent, SAR; Tyler to Bomford, 8 Jan. and 7 Mar. 1831, 7 Jan. 1832, Letters Recieved, OCO; Deyrup, *Arms Makers* (note 28), pp. 59–65.

37. U.S. *Statutes at Large* 3:615, 4:504; Falk, "Soldier-Technologist" (note 7), pp. 141–44.

38. Benet, *Ordnance Reports* (note 4), 1:235–36.

39. Talcott to Bomford, 15 Dec. 1832, "Inspection of Harpers Ferry Armory," 17–25 July 1835, Reports of Inspections of Arsenals and Depots, OCO.

40. Talcott, "Inspection of Harpers Ferry Armory," 3 Nov. 1836, Reports of Inspections of Arsenals and Depots, OCO.

41. See the file of Reports of Inspections of Arsenals and Depots, 1832–60, OCO.

42. *Regulations for the Government of the Ordnance Department* (Washington, D.C., 1834); William A. DeCaindry, *A Compilation of the Laws of the United States Relating to and Affecting the Ordnance Department* (Washington, D.C., 1872).

43. Wade, "Early Systems of Artillery" (note 11), pp. 142–43; Falk, "Soldier-Technologist" (note 7), pp. 286–87.

44. These developments are fully documented by Falk, "Soldier-Technologist" (note 7), pp. 228–85; and Birkhimer, *Historical Sketch of Artillery* (note 9), pp. 243–47, 262–65, 280–83. For information on the adaptation of woodworking machinery at the Springfield Armory to the manufacture of carriages at the Allegheny arsenal, see Thomas Blanchard, "An Estimate of the Cost of Machinery for making Gun Carriages for the U. States," 2 Feb. 1830, Letters Received, OCO.

45. Ibid.; Ezell, "Development of Artillery" (note 9), pp. 107–31.

46. Captain Benjamin Huger to Bomford, 28 Jan. 1841, Letters Received, OCO; U.S. *Statutes at Large* 5:512; Benet, *Ordnance Reports* (note 4), 2:35–36; Ezell, "Development of Artillery" (note 9), pp. 125–33. The results of Wade's work are documented in U.S. Ordnance Department, *Reports of Experiments on the Strength and Other Properties of Metals for Cannon* (Philadelphia, 1856), pp. 11–313. For his unpublished papers, see Reports of Experiments,

Tabular Reports of the Inspection of Cannon, Reports Relating to Tests Made on Cast Iron Cannon, and Reports on the Manufacture of Heavy Ordnance, OCO. Edwin Layton places Wade's investigations in larger historical context in his perceptive "Mirror-Image Twins: The Communities of Science and Technology in 19th-Century America," *Technology and Culture* 12 (1971):571–72.

47. Falk, "Soldier-Technologist" (note 7), pp. 302–4, 338–51, 354–60, 403–4, 526–27. Also see Mordecai's *Report of Experiments on Gunpowder, Made at Washington Arsenal, in 1843 and 1844* (Washington, D.C., 1845); and his *Second Report of Experiments on Gunpowder, Made at Washington Arsenal, in 1845, '47, and '48* (Washington, D.C., 1849).

48. Benet, *Ordnance Reports* (note 4), 2:60–75, 138–47, 185, 195–204; Ordnance Department, *Reports of Experiments on Metals for Cannon*, pp. 325–46. Walbach's original correspondence and reports are found in Reports Relating to Tests Made on Cast Iron Cannon and Reports on the Manufacture of Heavy Ordnance, OCO.

49. Benet, *Ordnance Reports* (note 4), 2:527–29; Ezell, "Development of Artillery" (note 9), pp. 133–37. For Morfit's reports, see Ordnance Department, *Reports of Experiments on Metals for Cannon*, pp. 371–428.

50. Falk, "Soldier-Technologist" (note 7), pp. 385–87. Rodman's investigations (1843–60) are detailed by Ezell, "Development of Artillery" (note 9), pp. 137–57. Also see Rodman, *Reports of Experiments on the Properties of Metals for Cannon, and the Qualities of Cannon Powder* (Boston, 1861).

51. Mordecai to Samuel Mordecai, 2 June 1861, quoted by Falk, "Soldier-Technologist" (note 7), pp. 593–94. Also see Falk, pp. 409–10.

52. Thomas P. Hughes presents a thoughtful analysis of these stages in "Inventors: The Problems They Choose, the Ideas They Have, and the Inventions They Make," *Technological Innovation: A Critical Review of Current Knowledge*, ed. Patrick Kelly and Melvin Kranzberg (San Francisco, 1978), pp. 166–82.

53. Smith, *Harpers Ferry Armory* (note 14), pp. 288–90; Cesari, "American Arms-Making" (note 29), pp. 92–131. Other examples are provided by David A. Hounshell, "The *System:* Theory and Practice," in this volume.

54. Nathan Rosenberg discusses the concept of technological convergence in his seminal article on "Technological Change in the Machine Tool Industry, 1840–1910," *Journal of Economic History* 23 (1963):423–30. For further documentation, see David A. Hounshell, "From the American System to Mass Production: The Development of Manufacturing Technology in the United States, 1850–1920" (Ph.D. diss., University of Delaware, 1978); Deyrup, *Arms Makers* (note 28), pp. 117–28, 146–59; Cesari, "American Arms-Making" (note 29), pp. 92–264; Battison, *Muskets to Mass Production* (note 15), pp. 16–19, 31; Roe, *English and American Tool Builders* (note 3), pp. 128–44, 164–215.

55. Cf. Paul Faler, "Cultural Aspects of the Industrial Revolution: Lynn, Massachusetts, Shoemakers and Industrial Morality," *Labor History* 15

(1974):367–94; Alan Dawley, *Class and Community: The Industrial Revolution in Lynn* (Cambridge, Mass., 1976); Smith, *Harpers Ferry Armory* (note 14); Thomas Dublin, "Technology, the Organization of Work, and Worker Solidarity: A View of Women in the Early New England Mills" (paper delivered at the Social Science History Association Conference, Ann Arbor, Michigan, 23 Oct. 1977); Richard P. Horwitz, *Anthropology Toward History: Culture and Work in a 19th-Century Maine Town* (Middletown, Conn., 1978); and Anthony F.C. Wallace, *Rockdale: The Growth of an American Village in the Early Industrial Revolution* (New York, 1978).

56. Undated armory regulations, ca. 1816, Letters Received, SAR; Lee to Levi Dart, 10 Oct. 1816, to Bomford, 2 Nov. 1816, Letters Sent, SAR.

57. Benet, *Ordnance Reports* (note 4), 1:401–3.

58. Major John Symington to Captain William Maynadier, 12 July 1849, Edward Lucas, Jr., to Bomford, 29 Aug. 1839, Letters Received, OCO; Talcott, "Inspection of the Harpers Ferry Armory," 17–25 July 1835, Reports of Inspections of Arsenals and Depots, OCO; Benjamin Moor to Major Rufus L. Baker, 5 May 1831, Letters Received, Allegheny Arsenal Records, Record Group 156, National Archives. For further documentation, see Smith, *Harpers Ferry Armory* (note 14), pp. 252-304.

59. *American State Papers: Military Affairs*, 2:553.

60. Bomford to Bell, 6 Apr. 1841, Bell to Bomford, 1 Apr. 1841, Major Henry K. Craig to Bomford, 1 Apr. 1841, William B. Calhoun to Bell, 1 Apr. 1841, Letters Received, OCO; Bomford to Craig, 2 Apr. 1841, Letters Sent, OCO; Benet, *Ordnance Reports* (note 4), 1:385–86, 389, 394–97.

61. Statement of John H. Strider to Secretary of War, 26 July 1848, Major John Symington to Talcott, 29 June 1846, 4 and 12 Aug. 1848, 12 May 1849, 21 May 1851, Letters Received, OCO; Benjamin Moor, "Objections to the Military Superintendencies of the National Armories," James H. Burton Papers, Yale University Archives; Charles Stearns, *The National Armories* (Springfield, 1852); Benet, *Ordnance Reports* (note 4), 1:431. Labor relations at the national armories are discussed by Deyrup, *Arms Makers* (note 28), pp. 164–66; and Smith, *Harpers Ferry Armory* (note 14), pp. 268–74, 298–300.

62. Talcott to John C. Spencer, 17 May 1842, Letters Received, OCO. Also see Colonel Henry K. Craig to secretary of war, 28 Oct. 1851, 2 Nov. 1852, 11 Nov. 1853, Letters Sent to the Secretary of War, OCO; Benet, *Ordnance Reports* (note 4), 1:395–97, 431, 501; 2:368–70, 436, 445, 532–37, 543–45.

63. "Manufacturing Industry of the State of New York," *Hunt's Merchants' Magazine* 15 (1846):371–72; Benet, *Ordnance Reports* (note 4), 1:50, 221, 431.

MEASURING TECHNIQUES AND MANUFACTURING PRACTICE

Paul Uselding

Over the past century or so various authors have undertaken the task of describing and reporting on specific machinery and processes—"hardware"—that form a part of the American system of manufactures. The history of this system has largely been oriented around specific events and artifacts. What has remained unclear is exactly what the disparate machines and processes contributed to the attainment of interchangeable manufacture. Until recently most scholars dealing with the topic of interchangeable manufacture have tended to treat such machines and processes as the heart of the matter.[1] In 1960 Professor Robert Woodbury provided a useful guide to the required elements in a system of interchangeable manufacture: (1) precision machine tools, (2) precision gauges or other instruments of measurement, (3) uniformly accepted standards of measurement, and (4) certain techniques of mechanical drawing.[2] To date, the bulk of historical work on the American system of manufactures has centered on point (1) above, the machine tools themselves. The present essay attempts to address Woodbury's points (2) and (3), measurement devices and standards.

The motivation for such an inquiry stems not only from its relative neglect in the historiography of interchangeable manufacture, but also from the belief that such words as "interchangeable," "accuracy," "precision," and many other terms encountered in this topic's technological history have quantitative and conceptual meanings that have remained latent and unexplored. There is more to the history of the American system of manufactures than machines and artifacts alone. We need to know more about the methods and devices for measuring the output of these machines and the dynamic interaction between mechanical processes and the standards of accuracy and precision embodied in measuring devices and techniques. In this limited question alone we have a classic chicken-and-egg problem. Which comes first in an evolving production system, the inherent accuracy of machine tools or techniques and devices which permit accurate measurement? This essay will argue that inherent machine accuracy—precision—is predicated upon the prior existence of accurate measurement devices and methods. The question of the dynamic interaction between machine and measurement systems cannot be addressed at the moment, since our knowledge of this subject has not progressed to a point at which we could even make intelligent guesses.

Throughout most of the nineteenth century, interchangeable manufacture was achieved by *precision*, not *accurate* measurement in the modern engineering sense. The essential feature of precision manufacture is exact duplication utilizing matched or common fixtures, tools, and size gauges. Accuracy of measurement in manufacture is the attainment of exact size relative to a standard measure. At the end of the nineteenth century, as production requirements became more stringent, as manufacturing facilities began to decentralize and become multinational, and as a number of other problems arose, accuracy, not precision, became the main principle of interchangeable manufacture. Also, tolerances—which, in this sense, are permissible variations in length around a mean or nominal dimension—first began to be used on engineering drawings around the turn of the twentieth century.[3] Today, most organizations concerned with engineering standards refer to three basic types of accuracy or tolerance: *size, form,* and *location.* Locational tolerance includes *concentricity, symmetry,* and *position.*[4] In the discussion which follows, emphasis is on size accuracy. Locational accuracy and form accuracy are vast subjects. Although they cannot be treated extensively here, it is well to point out that these important forms of mechanical accuracy are an integral part of the twentieth-century system of interchangeable manufacture.[5]

It is approximately correct to state that the "American system" of the nineteenth century was a precision system in which the principal type of accuracy improvement was size accuracy even though, most often, workpieces were produced to fit common fixtures, tools, and gauges and not to exact size relative to a universal standard of measurement. It will be shown that levels of size accuracy improved markedly as the nineteenth century progressed. Nevertheless, the accuracy of the American system of manufactures of the nineteenth century was not an inclusive accuracy of *size, form,* and *position* as is true of modern twentieth-century interchangeable manufacture. Regardless of the increasing levels of size accuracy attained, it remained the case that a good bit of fiddling and fitting was necessary to achieve workable mechanical assemblies in the absence of standards and precision measurement devices for achieving uniformity of form and position in mating workpieces.

Position tolerance includes such things as squareness and parallelism of surfaces, perpendicularity, angle accuracy, tooth or thread spacing, and hole location.[6] The principal early forms of measuring instruments relied on the multiplying lever embodied in such devices as the toolmaker's indicator or the dial gauge (dial test indicator). More recently, the practice has been to measure by sophisticated machines using electrical indicators and probes, and computers for coordinate analysis.

Accuracy of *form*, on the other hand, is determined by the method of production. It relates to the domain of the modern methods engineers who specify whether a surface is to be planed, ground, scraped, or lapped, depending on the quality of the finish required. Accuracy of form is another broad category of inquiry with important connections to the modern system of interchangeable manufacture. Form standards are important: because of their intimate connection with production methods they naturally raise questions about the relationships between the types of accuracy (*size, form,* and *location*) in systems of interchangeable manufacture. We obviously need to know a good deal more about the interactions between size accuracy and accuracies of form and location in order to understand more completely how the system of interchangeable manufacture evolved.

Beyond these considerations, let me suggest adding a fifth point to Professor Woodbury's list of the four required elements: techniques of managerial and administrative control. It is important to recognize that Woodbury's first two points embrace devices (hardware). It is customary to call such types of technology "embodied," in that the technical knowledge is literally embodied in the device or machine itself and is incapable of being separated in logic or fact from the specific form and substance of the machine or instrument. Items (3) and (4) on Woodbury's list are in the domain of "disembodied" technologies, i.e., knowledge or technique that is not congealed in a specific, tangible form. There are two forms of technology, embodied and disembodied, which must be dealt with in complete discussions of the American system of manufactures. Professor Alfred Chandler's essay in this volume deals with the important question of managerial and administrative control within a system of factory production.[7] The specifics of the ways in which measurement and inspection functions were handled in a managerial or administrative sense remain, nevertheless, questions for future investigation.

These points have been considered here because it is well to note the necessity of much additional research if we are to understand fully this complex entity we call the American system of manufactures. Think, if you will, about Professor Woodbury's four requisite elements for interchangeable manufacture augmented by the additional point mentioned above. By forming sensible pairwise hypotheses we have a large number of research questions. For example, considering points (1) and (4), how did the evolution of mechanical drawing and other means of communicating technical information affect the design and capability of machine tools? Or, looking at points (3) and (5), what administrative techniques and procedures were used to impose "accepted standards of measurement"? What effect did these administrative procedures, coupled with measurement and inspection techniques, have upon product cost and quality? When we move beyond research questions formed by pairwise considerations from this list, the horizons for research expand enormously, as does our potential for understanding the full implications and set of causal factors in this technological episode. The scope of this research agenda is even more imposing when it is realized that such questions are all internal to the process of interchangeable manufacture itself. Other essays in this volume touch upon the external implications of the American system of manufactures, especially those by Daniel Nelson (impact on workers) and Neil Harris (consumer behavior and motivation). These provide the reader with an entirely different avenue of approach to the topic.

For the present, however, most of our work remains at the level of exploration of discrete topics. Because the machine tools themselves have received the most attention to date, this essay undertakes to "move down Woodbury's list," so to speak, to initiate the exploration of another important, if discrete, element in the requisite list for the attainment of interchangeable manufacture.

Measuring Techniques and "Hardware" Historiography

If the contributors to this volume have not taken up the "hardware" question directly, it is very likely because of a mutual feeling that the areas where considerable future progress is to be made will be on topics other than Woodbury's first point—the machine tools themselves. The accumulation of new hardware knowledge is a slow and difficult pursuit, resting on chancy retrievals from public and private archives and manuscript collections, patent files, and artifacts collections. It is not the kind of work that can be ordered up "on demand." A new finding, limited to a specific process or machine at a specific date, may require years of investigation and inquiry.

Nevertheless, because a sound background in the literature of the American system of manufactures dealing specifically with machines and process is assumed in the level of treatment given here and in several other essays in this volume, I will set out some of the main contributions to the "hardware literature." These studies form the basis on which rests much of our discussion of the system.

Two bedrock studies, consulted by almost every student of the subject, are Charles Fitch's "Report on the Manufactures of Interchangeable Mechanism" and Joseph Roe's *English and American Tool Builders*.[8] Fitch's work, published in 1883 as a supplementary report in the tenth census, consists of detailed reports of machines, processes, and shop practices in such key industries as firearms and sewing machines, and in several others. It is liberally illustrated and incorporates details of technological practices not only in the decade of the 1870s but also descriptions based on interviews with and recollections of machinists familiar with contemporary practices as early as the 1830s. Joseph Roe's book might be termed a technical biography. It gives important information on the lives and accomplishments of prominent English and American machine-tool builders and engineers and provides family histories of important nineteenth-century metalworking establishments—showing the relationships of firms to each other over time along with descriptions of the transfer of personnel and techniques that attended these changes in ownership and firm name.

An excellent, if brief, overview of machine-tool developments between 1450 and 1850 is contained in Abbott Payson Usher's *A History of Mechanical Inventions*.[9] In addition, the work of Nathan Rosenberg, Wayne Broehl, and Harless Wagoner provides important material on the general nature of the American machine-tool industry in various periods.[10] Rosenberg's study isolates the reasons for the early emergence of a standardized machine-tool industry in America. The phenomenon of *technological convergence*—i.e., the interaction of relatively unsophisticated machine-buying demands from a burgeoning industrial structure with the engineering knowledge located in the machine-building firms—is held to account for the emergence of standard machine tools capable of adaptation to a wide variety of technically related uses. Wayne Broehl's work focuses on the development of specific machine-tool firms in a specific location, Springfield, Vermont. Harless Wagoner's study is a broad overview of the American machine-tool industry between 1900 and 1950, dealing mainly with managerial and business issues.

Robert Woodbury's quaternity of monographs on the history of specific machine tools, published between 1958 and 1961, are fundamental contributions to the history of gear cutting, grinding, milling, and turning.[11] They are soundly researched and written in an interesting and comprehensible manner. Handsomely illustrated, they also provide excellent coverage of early important source material on their respective topics.

Detailed studies of individual machines or firms are found in the work of Edwin Battison, Felicia Deyrup, Merritt Roe Smith, and myself.[12] Battison has carefully examined surviving firearms manufactured by Eli Whitney in the early nineteenth century and concluded that, contrary to the legend of Whitney's priority as inventor of the milling machine, true milling was not practiced in his shop. Deyrup's study of the Springfield Armory contains a wealth of information on early technology at this important federal center, which was pivotal in the diffusion of new techniques. Roe Smith's work provides solid evidence of the ways in which mechanical skill was transmitted between the government and private arms contractors by the migration of skilled mechanics. In addition, his recent book, *Harpers Ferry Armory and the New Technology*, contains several meticulously researched chapters on early machine technology at Harpers Ferry.[13] My own work concerns the transfer and diffusion of techniques for forging and rolling metal, tracing the development of mechanized barrel-rolling at Springfield Armory as initiated by an immigrant Scot, Henry Burden. The importance of experience in forging practice in one industry (axes) for subsequent innovation in another (firearms) is also illustrated in my examination of Elisha K. Root's career as a mechanic. Most of these studies contain detailed descriptions and illustrations of significant machines.

Frequent reference is made by contemporary historians of technology to the work of engineers and technical writers published between approximately 1900 and 1925. Much of this work appeared in *American Machinist*, a technical periodical, and the reader will do well to watch for the names of authors such as Luther Burlingame, E.A. Dixie, Guy Hubbard, and E. G. Parkhurst.[14]

Two books have done much to stimulate research by economic historians in this area: H. J. Habakkuk's *American and British Technology in the Nineteenth Century* and David Landes's *The Unbound Prometheus*.[15] Habakkuk's 1962 book sparked a great deal of interest in the economic forces influencing the relatively early development of the American system of manufactures in this country and the slowness of the British to adopt interchangeable manufacture and the techniques of mass production generally. Landes's work, while not concerned strictly with American developments, is mentioned here because its first two chapters provide a definitive treatment of the development of the textile industry in England. It was the diffusion of knowledge about mechanized textile production to America that provided a necessary condition for the growth of that great "market for machine tools." Elsewhere I have published an extensive review of the economic-historical literature (mainly journal articles) published since 1960 on various topics such as labor scarcity, choice of technique and innovation, transfer and diffusion, productivity measure, biased technical change, and learning by doing.[16]

Having outlined the point at which topic of measurement standards and measuring instrumentation fits into the historiography of the American system of manufactures, we are now in a position to turn to an exploration of several critical questions. First, can more precise conceptual and quantitative meanings be given to such frequently used terms as "accuracy" and "precision"? Second, what was the relationship of earlier techniques of machine building and construction by skilled mechanicians using relatively simple tools to the later system of machine production in volume attaining limited interchangeability? What individuals contributed to the attainment of measurement standards systems? What were the principal devices and instruments of measurement that contributed to the higher standards of accuracy as the nineteenth century progressed? What levels of statistical tolerance were achievable at various dates? To these and other questions we now turn our attention.

The "Making" and "Manufacturing" Systems

In the early decades of the nineteenth century there coexisted two contrasting systems of machine production. Employing a distinction used by Charles Babbage in his celebrated work *On the Economy of Machinery and Manufacture*, published in 1832, these may be termed the *making* system and the *manufacturing* system.[17] In the *making* system, machines or machine-fabricated components are made one at a time, or in limited quantities. Under the *making* system the methods of production are not greatly affected by the number of units produced. In the *manufacturing* system, which embraces large-scale production and interchangeable manufacture, the production methods employed are a direct function of the number of units produced. Of the external stimuli to the emergence of the American system of manufactures, the existence of a sufficient demand for standardized, machine-made products to generate the volume required for alteration in production method stands in an important place.

Babbage's distinction, while simple and direct, also has the nice virtue of helping to parcel out the right amount of "credit" to the various sovereign nations and their native inventors. To England we owe the origination of the making system and its attendant machine tools. The manufacturing system, and the machine tools that are among its hallmarks, is American. The English making system originated the standard machine tools—the lathe, planer, shaping machine, slotting machine, boring mill, drilling machine, and gear-cutting machine. The American mutations of these original tools entailed defining elements of the manufacturing system such as the milling machine and profiling machine, automatic gear cutter, turret lathe, automatic screw machine, grinding machine, multiple spindle drill, and the panoply of special machines, gauges, fixtures, and jigs.

It is an obvious oversimplification to state, as has been done here, that the mechanical universe of the nineteenth century revolved around the twin axes of England and America, so perhaps we should substitute "Europe" for England. Moreover, it should not be inferred that Babbage's categories imply a mutual exclusivity in practice. It was often the case that the making and manufacturing systems existed side by side in the same factory, jointly contributing to the creation of the final product.

The manufacturing system could not have been developed without the prior existence of the making system. Moreover, it is ironic to contemplate, but I believe essentially correct to assert, that the interaction between these two production systems has been asymmetric—the manufacturing system on balance being a net borrower from the making system. It is often assumed that the sine qua non of the American system of manufactures is mechanical precision. Such precision arises from the inherent capability of the machines employed and the methods of controlling the dimensions of machined workpieces. While this is undoubtedly true, it is also the case that mechanical precision and accurate measurement were not the exclusive province of the manufacturing system.

Measurement in the English Making System

About 1800 Henry Maudslay, the great English toolbuilder doubly devoted to accuracy in the creation of true planes and in perfect screw

Replica of Henry Maudslay's "Lord Chancellor," ca. 1800. (From the collections of the Division of Mechanical and Civil Engineering, National Museum of American History; Smithsonian Institution Neg. No. 58606.)

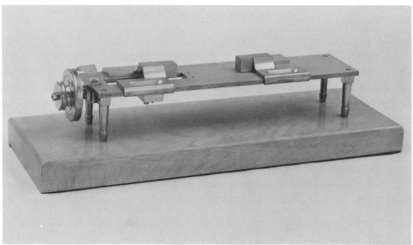

threads, built a measuring instrument, the famous "Lord Chancellor." It was an amazingly accurate device, even by modern standards. Employing a master screw, the "Lord Chancellor" had 100 threads per inch (approximately four threads per millimeter) and could detect extremely minute differences in length. James Nasmyth, a Maudslay disciple, described the instrument:

> Not only absolute measure would be obtained by this means, but also the amount of minute differences could be ascertained with a degree of exactness that went quite beyond all the requirements of engineering mechanism; such, for instance, as the thousandth part of an inch [0.0254 mm]. It might also have been divided so far as millionth part of an inch [0.000025 mm] but these infinitesimal fractions have really nothing to do with effective machinery that comes forth from the workshops, and merely show the mastery we possess over materials and mechanical forms.[18]

By midcentury another Maudslay disciple, Joseph Whitworth, had carried mechanical precision to a new level within the English making system. In 1856 Whitworth delivered a presidential address to the Institution of Mechanical Engineers and emphasized "the vast importance of attending to the two great elements in constructive mechanics, namely, a *true plane* and *power of measurement*. The latter cannot be obtained without the former, which is, therefore, a primary importance... all excellence in workmanship depends upon it."[19] In 1855 Whitworth exhibited a "measuring machine" based on end measurement, which, it was claimed, could measure differences of one-millionth of an inch (0.000025 mm). Whitworth was also quite clear on the advantages of end measurement:

> We have in this mode of measurement all the accuracy we can desire; and we find in practice in the workshop that it is easier to work to the ten-thousandths of an inch [0.0025mm] from standards of end measurements, than to one-hundredth of an inch [0.25mm] from lines on a two-foot rule [610 mm]. In all cases of fitting, end measure of length should be used, instead of lines.[20]

These examples should serve to buttress the point that the making system of the great English toolbuilders embodied a high degree of machine and measurement precision. Taking these examples at face value, the English making system, at the opening of the nineteenth century, was capable of producing machine work to tolerances of one-thousandth of an inch. By midcentury, Whitworth's own account claims that exactness to a ten-thousandth of an inch was within the range of practice in his shop.

In America, Ambrose Webster, whose career in mechanics included extensive experience at the Springfield Armory and the American Watch Company between the 1840s and 1860s, had this to say in 1894: "Among the many things which have led to the development of the present high skill and precision in machine shops, I consider that steel scales and micrometer calipers stand in the front rank."[21] Since high-order precision was present both in the English *making* system and the American *manufacturing* system by midcentury, it remains to be explained how measurement accuracy can be claimed to "stand in the front rank" of characteristics of the "American system" of interchangeable manufacture. Put another way, how could English toolbuilders have developed high quality machine tools so rapidly before 1850 without interchangeability? And how could American manufacturers have achieved interchangeability by midcentury without the parallel development of uniform measurement standards? Without being able to adduce anything approaching conclusive proof, let me suggest that the resolution of this seeming paradox may be found in the way measurement was applied in the two systems. The tendency among English mechanics and engineers was to make parts to *fit* (or to mate to each other) rather than to *dimensions*. While Americans had by no means begun to approach a set of uniform standards by midcentury, in-shop standardization of measure and method was occurring in various branches of manufacture with increasing frequency, and it was not seldom the case that interchangeability was achieved between the parts of two or more manufactures.

At a less general level of explanation there is another characteristic of the way measurement is applied in the *making* system that distinguishes it from the *manufacturing* system in the American case. In the United States line measures or graduated scales were the principal means of measurement in the making system, whereas in the manufacturing system the main reliance was on end measurement and limit gauges. End measurement typically employs devices such as rod gauges, plug and ring gauges, and solid caliper ("snap") gauges. In the American manufacturing system such end measures were applied directly to the workpiece for "sizing." While the shop practice of Joseph Whitworth can hardly be referred to as typical of English practice generally, it is instructive to note the difference in *application* of end measurement in Whitworth's shop. For example, Whitworth's plug and ring gauges were not developed for direct use in gauging parts, and Whitworth did not use them in that fashion. Rather he substituted his plug and ring gauges for the customary graduated scales, with the workman transferring his caliper measurement from the gauges. Ideally, the dimension of the workpiece was then a duplication of the size of the gauge through the intermediation of the workman's caliper measurement.

Each type of measurement has its own specific advantages and disadvantages. In end measurement the sense of touch is used, whereas with line measures the sense of sight is foremost. Within the limit of its length and graduations the line measure gives all sizes while the end measure establishes only a single fixed dimension. Hence, end measurement requires a large number of special-purpose gauges and for this reason is far more expensive to use than line measurement. In addition, line measures do not lose accuracy in use, while end measures are subject to wear or "going-out-of-gauge," which further tends to increase the cost of maintaining the accuracy of an inspection system based upon end-measurement principles.

Questions of cost aside, the problem with line measures is that in transferring lengths from scales by positioning the legs of the caliper upon the graduation marks, "sight errors" attributable to the workman can occur. Additional errors can enter during the process of transferring the caliper setting to the workpiece and scribing the measure. The recognition of such errors in "sizing" from graduated scales led Bodmer, about 1840, to design the first plug and ring gauges. Undoubtedly, similar considerations entered into the development of Whitworth's famous plug and ring gauges. It is interesting to note in passing that tests with a surviving set of Whitworth plugs and rings, imported into the United States in 1857 and now in the possession of the American Society of Mechanical Engineers, showed *real* accuracy well below Whitworth's claimed or *nominal* accuracy. This was the case even though the gauges themselves were not worn. Similar results were obtained from a set of Whitworth gauges tested by the Engineering Standards Committee of Great Britain.

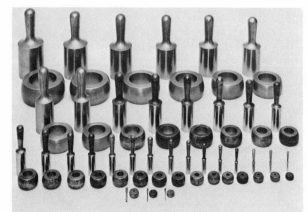

A set of Whitworth plug and ring gauges, 1870. (On loan to the Division of Mechanical and Civil Engineering, National Museum of American History, from the Brown and Sharpe Manufacturing Co.; Smithsonian Institution Neg. No. 81–202.)

Results of this kind should make us aware of the important distinction between *real* and *nominal* measurement accuracy. James Watt's shop micrometer, now in the Science Museum at South Kensington, has eighteen threads per inch and its dial is divided into 100 parts, making the nominal value of its smallest unit of measure 1/1800th of an inch. While an instrument such as the Watt micrometer shows a clear conception of accuracy and did provide, in principle, a means for its measurement, its *real* was not equal to its *nominal* accuracy, as it was not possible in Watt's time to make screws with this *real* degree of accuracy.

Whitworth's "millionth measuring machine" would now be called a comparator, since its purpose was to compare different pieces and determine their differences, not to measure absolute length. That being so, we may leave this device out of consideration in considering the influence of accurate measurement in evolving precision within the making system. (Whitworth's comparator has been called a machine which attempts to "weigh a shadow.") Whatever the divergence between real and nominal accuracy within the Whitworth system of measurement, by the seventh decade of the nineteenth century Whitworth had made the thousandth of an inch a common measure, the ten-thousandth a tangible reality, and the hundred-thousandth something more than a philosophical goal.

American Developments: Shop Standards and Practices

Having explored the standards of accuracy for measurement within the *making* system as practiced in England, we now turn to the American case, hoping to find further clues to the puzzling way precision measurement enters the two contrasting production systems. Charles Hildreth, superintendent of the Lowell machine shop, provided a good insight into early nineteenth-century practices in an article appearing in *American Machinist* in 1894:

> The original unit of length, adopted by the Lowell Machine Shop in 1824 for use in the construction of machinery, is the English foot, according to a brass standard made by Cary of London, and preserved with care in the safe of the proprietors of the locks and canals on Merrimack River. This scale is undated, but has marked upon it "Cary, London, Fahrenheit, 62 degrees.". . . From 1824 to 1835 it was impracticable for the workmen employed at these shops to readily procure scales and rules for measuring lengths, which scales and rules were in exact accord with one another. One of our superior mechanics, Samuel Brown, made in 1835 a standard screw for use in graduating measuring scales. This screw was made to agree with the standard brass scale of Cary, of London, and it took nearly a year to perfect it. Mr. Brown then made

two short duplicate standard screws for use in graduating small scales. . . . The Lowell Machine Shop made standard scales long before they were undertaken to be manufactured [by Darling and Bailey] in Bangor.[22]

This passage reveals that substantial gains could be obtained within the system of line measurement, i.e., by converting from boxwood scales to steel scales graduated from a standard screw. Moreover, line and end measures coexisted in America within the manufacturing system of the famed Springfield Armory. As Ambrose Webster attests:

The machinist of today [1894]. . . has little conception of the difficulties of making accurate measurements which confronted the machinist of 1840.

When [I] commenced [my] term of four years' apprenticeship there were no steel scales in the market, and we were obliged to set our calipers to the proper dimensions by laying them on ordinary boxwood rules, trying to have the ends of the caliper coincide with the division lines. It was utterly impossible, of course, to secure uniform measurements in this way, and even as late as 1856 there were, in the machine shop of the American Watch Company, four different sizes of "standard" ½-inch taps.[23]

Webster goes on to state that in 1851 he fashioned a handmade steel scale for use in the Springfield Armory. From the accompanying illustration it can be seen that this scale divided the inch into sixty-four parts. He indicates that such scales were in use at the armory at this time but does not provide information on how widespread their use was among Springfield mechanics.

In 1850 Joseph Brown of Brown & Sharpe began to manufacture steel rules and scales and vernier calipers. Brown foresaw the difficulties of machining an accurate screw for the purpose of graduating the scales, so he developed his famous graduating engine, whose specific characteristics remained a carefully kept secret. In 1852 Samuel Darling and Edward Bailey also commenced the manufacture of steel rules in Bangor, Maine. This partnership manufactured the famous "Bangor Scales" until 1866, when Darling formed a partnership with Joseph Brown and Lucian Sharpe. The appearance of these commercial establishments in the early 1850s indicated the presence of significant demand for steel scales and rules in the machine shops and factories of America. Prior to this time Springfield Armory was using inspection gauges (end measures). E. A. Dixie, writing in *American Machinist*, discusses some of the filing jigs and gauges used there between 1840 and the end of the century.[24] Dixie shows an illustration of a breech-screw inspector's gauge—perhaps, more correctly, a flat steel template—believed to be in use around 1839 in

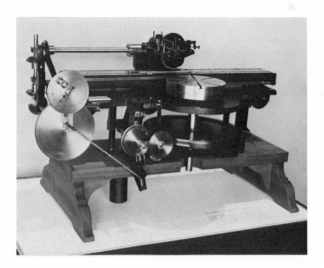

Brown's Model 3 Graduating Engine, 1859. (On loan to the Division of Mechanical and Civil Engineering, National Museum of American History, from the Brown and Sharpe Manufacturing Co.; Smithsonian Institution Neg. No. 58488.)

addition to a number of filing jigs used in this period and subsequently. Dixie mentions that a common dimensional tolerance at the armory was ±0.01 of an inch. By the end of the nineteenth century many parts of the Springfield rifle demanded dimensional limits of ±0.001 inch. If these representations may be taken at face value, they provide a benchmark of the improvement in real accuracy in American manufacture during the last half of the nineteenth century. A gauge identified as "U.S. Swivel punch gauge for Clasp bands Mod. 1861" is stated to have a size-limit difference of 0.005 inch.[25] While such evidence is hardly conclusive in terms of pinning down specific accuracy ranges for machine work within the evolving American manufacturing system, it does serve to substantiate the notion that the precision was not of a higher order than that obtainable within the English making system.

As long as the American system remained oriented toward the use of special-purpose gauges, it remained expensive. The expense of creating special-purpose measuring instruments along with special-purpose machine tools could only be justified if production volumes were sufficiently large to bear the pro rata expense thereby incurred. As long as this was the case, interchangeability could be obtained only in those branches of manufacture in which production volumes were large. For this reason, it is doubtful that the diffusion of practical interchangeability at the shop level—or in factories in which small-batch production was the rule—had proceeded very far in America by the close of the Civil War. In order for the newly created conceptions of measurement to ramify widely in the American industrial sector, it was necessary for accurate measuring instruments to be made cheaply *and* to be general-purpose in nature.

The revolutionary device, heralding the dawn of an era of widely diffused mechanical precision in America, was the micrometer caliper. The micrometer was hardly a new conception. The Adler Planetarium in Chicago has two instruments in its collection, one by Christopher Treschler of Dresden dated 1572, and the second dated 1727 and signed by Carolus Joseph Merula of Milan. The earliest known micrometers date from the sixteenth century, with such men as Gascoigne, Horrocks, and Crabtree having used them for astronomical purposes about 1640.

Although the date when the micrometer was first used in making or manufacturing is not generally agreed upon, the origin of the American micrometer is an interesting story. In 1867 the Bridgeport Brass Company had a lot of sheet brass returned from the Union Metallic Cartridge Company for being "out of gauge." The sheets were made to the gauge in possession of the manufacturer but did not agree with the gauge used by the customer. When the two gauges were tested against a third standard, no two agreed with each other. All three gauges were of the commonly used slotted type. The Bridgeport Brass Company thereupon devised a micrometer type of gauge and subsequently approached the firm of J.R. Brown & Sharpe with the idea of having it manufacture sheet-metal gauges upon the micrometer principle. At the time this gauge was not thought practical because the intersecting spiral and straight lines scribed onto the barrel precluded the use of numerals; hence the instrument was difficult to read. In August 1867 Joseph Brown and Lucian Sharpe visited the Paris Exhibition, where they saw the "screw caliper" patented in 1848 by Jean Laurent Palmer and known as the "Système Palmer." This caused them to recognize the value and possibilities inherent in the micrometer.

Système Palmer, screw caliper patented in 1848 by Jean Laurent Palmer. (From the collections of the Division of Mechanical and Civil Engineering, National Museum of American History; Smithsonian Institution Neg. No. 41701–B.)

In Palmer's micrometer the graduations were placed on the stationary barrel, the other set on the revolving thimble—an arrangement which permitted stamping of figures to facilitate reading. In Joseph Brown's own words, "As a gauge was wanted for measuring sheet metal, we adopted Palmer's plan of division and the Bridgeport man's size of gauge, adding the clamp for tightening the screw and adjusting screw for compensating the wear of the points where the metal is measured and thus produced our 'Pocket Sheet Metal Gage.'"[26] This "Pocket Sheet Metal Gage" was marketed by Brown & Sharpe until 1877, when they introduced the term "micrometer caliper" into their catalog.

Prior to the introduction of the micrometer in America in the late 1860s, fixed gauges, calipers, and rules were the only measuring instruments available on the shop floor in establishments engaged in interchangeable manufacture. More often than not, fits were obtained by trial and error. Fixed gauges were reasonably accurate but expensive to use; line measures were cheap but relatively inaccurate, seldom capable of measuring to within ±0.01 of an inch.[27] Compared to these earlier devices used for inspection in the American system of manufactures, the micrometer was accurate and inexpensive enough to enable every machinist to own one. The micrometer made it possible for workmen to measure the parts they made with reasonable assurance that they were to size within 0.001 of an inch. The micrometer made the "thousandths" a generally attainable standard of accuracy in American machine practice in the 1870s. Today, a micrometer in the hands of a trained machinist can be used to measure with reliability within five ten-thousandths of an inch (0.0005). Its inherent accuracy is usually between fifty-millionths to one ten-thousandths of an inch (0.000050 to 0.0001).

A Brown and Sharpe advertisement of 1868. (From the collections of the Division of Mechanical and Civil Engineering, National Museum of American History; Smithsonian Institution Neg. No. 81–187.)

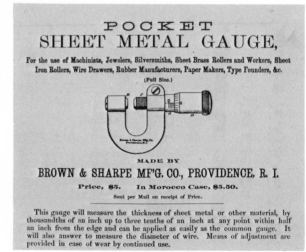

POCKET
SHEET METAL GAUGE,

For the use of Machinists, Jewelers, Silversmiths, Sheet Brass Rollers and Workers, Sheet Iron Rollers, Wire Drawers, Rubber Manufacturers, Paper Makers, Type Founders, &c.

(Full Size.)

MADE BY
BROWN & SHARPE MF'G. CO., PROVIDENCE, R. I.

Price, $5.   In Morocco Case, $5.50.

Sent per Mail on receipt of Price.

This gauge will measure the thickness of sheet metal or other material, by thousandths of an inch up to three tenths of an inch at any point within half an inch from the edge and can be applied as easily as the common gauge. It will also answer to measure the diameter of wire. Means of adjustment are provided in case of wear by continued use.

The next phase in the development of precision measurement came between 1874 and 1882, when such men as Beale, Sweet, and Rogers and Bond ushered in a new era of measurement accuracy and dimensional standardization. Their contribution to the evolving system of manufacturing was to design and build measuring machines referenced to a standard yard. These measuring devices were used to gauge gauges. This periodic inspection of fixed gauges and line measures by measuring machines referenced to standard yard-bars represents a new level of quality control within the system of manufacture. Like all end measures, fixed gauges are subject to wear and deterioration, i.e., "going-out-of-gauge." When this happens the accumulation of errors in mating parts results in a loss of time and hence efficiency in assembly. The ease of assembly of mating parts is one of the hallmarks of the system of interchangeable manufacture. Periodic inspection of fixed gauges prevents high scrap and rework costs, loss of assembly economies, and loss of product reliability. For this reason the development of "master gauges" to gauge the gauges used by workmen represents a higher plateau of precision and uniformity. By the early 1880s the American manufacturing system had made the dimensional tolerance of ±0.001 of an inch generally attainable, and the "tenth" (0.0001 of an inch) an achievable production standard in a few firms such as Pratt & Whitney and Brown & Sharpe.

Professor John E. Sweet of Cornell University, a founder of A.S.M.E., conceived the idea of manufacturing standard "solid calipers" (fixed gauges) of sheet steel. They were to be as accurate as the gauges of Whitworth and yet be cheap enough to bring them into everyday use. To insure the accuracy of these solid calipers, Sweet built his measuring machine in the machine shop of the Sibley College of the Mechanic Arts at Cornell in the year 1874. At this time the authority of Whitworth's dictum regarding the superiority of end measures had not yet been overturned, hence Sweet's measuring machine was based on end measurement. In 1875 its accuracy was tested and found to be within 1/4480th of an inch (0.00022 of an inch, or 2.2 "tenths").[28] The influence of Sweet's machine was limited, but the Syracuse Twist Drill Co. did adopt his designs and became a notable manufacturer of precision gauges. Frederick A. Halsey, a student of Sweet's, was familiar with his pioneer work on gauges. Halsey edited an engineering periodical and was a recognized authority on machine-shop practice. In his book *Methods of Machine Shop Work* published in 1914, he comments: "The beginning of accurate shop measurements in the United States, was by Dr. John E. Sweet at Cornell University...."[29]

In 1878 Oscar J. Beale, who was chief inspector and foreman of the gear

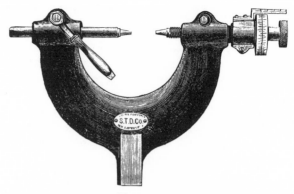

John E. Sweet's measuring machine, 1874. (From American Machinist, *8 Dec. 1883; Smithsonian Institution Neg. No. 81–191.)*

SWEET'S MEASURING MACHINE.

department at Brown & Sharpe, made a measuring machine which served those shops for many years as an original standard, as well as comparator. The significance of Beale's contribution is that it broke with the tradition of end measurement. Employing a microscope referenced to a standard yard, its measuring wheel was graduated to read to a ten-thousandths of an inch (0.0001), and its vernier read to a hundred-thousandths of an inch (0.00001).[30]

The ultimate American achievement in measurement in the nineteenth century was the work of Professor William A. Rogers, an astronomer at

Rogers-Bond Universal Comparator (early 1880s), built by Pratt and Whitney. (From the collections of the Division of Mechanical and Civil Engineering, National Museum of American History; Smithsonian Institution Neg. No. 73-6437.)

Harvard University, and George M. Bond of Pratt & Whitney, which involved the construction of standard reference bars and in 1882 resulted in the Rogers-Bond comparator for the accurate subdivision of these bars. The Rogers-Bond comparator was used not only to inspect fixed gauges in the Pratt & Whitney works, but, with its accuracy of one-millionth of an inch (0.000001), it was instrumental in the development of commercial measuring machines used for determinations of length to the hundred-thousandth of an inch (0.00001). These commercial measuring machines of Pratt & Whitney's, based on the Rogers-Bond comparator, were sold throughout the world in the late nineteenth and early twentieth centuries. Rogers and Bond demonstrated the practicability of originating exact original standards by the method of line measurement or, as they termed it, "calipering under a microscope." In 1884 George Bond stated: "It is possible and entirely practical to produce gauges so nearly alike by this means, that a variation between any two, of even one forty or one fifty-thousandth of an inch can be discovered."[31]

By the close of the nineteenth century Carl Edvard Johansson of Sweden had embarked upon the conception and execution of a combination gauge block system. Today, the term "jo-block" is synonymous with precise measurement in machine shops throughout the world. Johansson is claimed by his biographer Torsten K. W. Althin to have been inspired by the work of Christopher Polhem and the mathematical possibilities of his combination measuring sticks or "Polhem sticks." To attain this goal Johansson employed a number of gauge blocks of carefully calibrated sizes arranged in series to subdivide any measurement range. In the Johansson system, for example, 102 gauge blocks can be arranged to make 20,000 different combinations of measurement in the metric or English system. Johansson's work, begun in 1896, had advanced by 1908 to the point where any systematic combination of his gauge blocks up to 100 mm in length represented an accuracy of plus or minus a thousandth part of a millimeter, a standard of accuracy seldom met in the workshop. Working on a converted sewing machine, Johansson had adapted an improved method of grinding first known in 1740 and published by Gabriel Polhem, the son of Christopher Polhem. It followed Christopher Polhem's method for grinding cast-iron rolls, employing cottonseed oil and sieved emery as the grinding material. Johansson's grinding wheel was made of cast iron, while Gabriel Polhem's wheel had been based on a turned steel roll rotating in a mantle made of tin. The use of metal as a grinding medium is a precursor of the method now known as lapping, a method for fine-finishing shapes and surfaces in hardened pieces of metal. Another key element in the Johansson system was stress relief to eliminate temperature variation of the blocks. His conception of the principles involved is amusing and bears repetition:

> The molecules in the steel are like little children. When they get warm, they dance and jump around. When they get cold, they quiet down and go to sleep. After you wake them up and put them to sleep enough times they get tired, and the last time they stay asleep. In this manner the gage blocks are made stable and "stay." [32]

In actual practice, stress-relieving was performed by alternately deep-freezing and thawing the blocks to room temperature.

In 1903 the BIPM had found Johansson's blocks to be accurate to within one-tenth of a micron or four-millionths of an inch. By 1907 the improvements in finishing the blocks enabled them to be "wrung" together with a force of seventeen atmospheres, and by 1916 with up to thirty-three atmospheres. In 1914, jo-blocks were offered for sale in America with a price tag of $900 for a set. Upon receiving a set at the Delco plant in Dayton, Ohio, Charles Kettering observed:

> Gentlemen, this is the finest measuring tool in the whole world. The sooner we learn to use it, so much the sooner we can stop "fitting" parts together and simply "put" them together. The day is now not far off when every toolmaker will have to have a set of these blocks at his elbow, or he just won't know what the hell he is doing. Let's buy ten sets and get them out in the tool room and inspection departments. [33]

During World War I the German submarine blockade in the Atlantic necessitated a dramatic smuggling effort to maintain a supply of jo-blocks for American industry. At the war's end Henry Ford observed: "This must not happen again; we must have the secret methods within our country's boundaries. I will buy the manufacturing rights." [34]

A set of Johansson gauge blocks, 1934. (From the collections of the Division of Mechanical and Civil Engineering, National Museum of American History; Smithsonian Institution Neg. No. P64389–A.)

In 1923 Johansson became associated with Henry Ford, applying his skill and system to the mass production of the automobile. From that time forward the precision of the Johansson combination gauge blocks underwrote interchangeability in American automobile production. The American system of manufactures was no longer uniquely "American," but interchangeable manufacture had reached out to touch the daily lives of nearly every American family. Automobiles were the vanguard of an endless procession of familiar consumer durables—based on precision manufacturing and accurate measurement—that began to define a uniquely American lifestyle.

Notes

1. A notable exception to this generalization is the contribution by Merritt Roe Smith which appears in this volume. Smith's emphasis on the importance of the imposed goal of uniformity by U.S. Ordnance officers underlines the central role of the value structure and goal orientation of the administrative hierarchy in which technical choices are made for the attainment of specific technological outcomes.

2. Robert S. Woodbury, "The Legend of Eli Whitney and Interchangeable Parts," *Technology and Culture* 1 (1960): 235–54. This article has been reprinted in *Essays in American Economic History*, ed. A.W. Coats and Ross M. Robertson (London, 1969), pp. 49-62; and in *Technology and Culture*, ed. Melvin Kranzberg and William H. Davenport (New York, 1972), pp. 318-36.

3. Earle Buckingham, *Principles of Interchangeable Manufacture* (New York, 1941); and S.B. Littauer, "Development of Statistical Quality Control in the United States," *American Statistician* 4 (1950): 14-20.

4. John V. Liggett, *Fundamentals of Position Tolerance* (Dearborn, Mich., 1970), p. 3. A tolerance is the permissible variation in a specified relation, such as size, form, or position.

5. Only a partial incursion into the subject of positional tolerance is possible. The British led the way during the 1930s and 1940s in the development of a formal system for position tolerancing, although evidence of work on such systems can be seen in the work of Earle Buckingham in the early 1920s and in various industrial standards on gauging, such as that published by the Chevrolet Division of General Motors in 1940. See ibid., pp. 4-5; and Buckingham, *Principles of Interchangeable Manufacture* (note 3), passim. The early American efforts were defined gauge systems, which were essentially position tolerances expressed in terms of gauge requirements as opposed to a formal system of position tolerances. The point of mentioning this is not to enter into the domain of national competition over priority or to give an inclusive account of the history of the determinants of modern accuracy, but

rather to indicate something of the importance of this augmented view of measurement accuracy.

6. Position tolerance is the distance between the fundamental surface of a workpiece and its allowable boundary.

7. In addition to Alfred Chandler's essay in this volume, another, more specific approach to understanding the role of such factors as the accounting system, plant layout, piecework arrangements, and division of managerial control and responsibility at Springfield Armory before 1850 may be found in Paul Uselding, "An Early Chapter in the Evolution of American Industrial Management," in *Business Enterprise and Economic Change*, ed. Louis P. Cain and Paul Uselding (Kent, Ohio, 1973), pp. 51-84.

8. Charles H. Fitch, "Report on the Manufactures of Interchangeable Mechanism," in *Report on the Manufactures of the United States at the Tenth Census* (Washington, D.C., 1883), pp. 607-707; and Joseph Wickham Roe, *English and American Tool Builders* (New York, 1916).

9. Abbott Payson Usher, *A History of Mechanical Inventions* (Cambridge, 1929 and 1954).

10. Nathan Rosenberg, "Technological Change in the Machine Tool Industry, 1840-1910," *Journal of Economic History* 23 (1963): 414-43; Wayne G. Broehl, *Precision Valley: The Machine Tool Companies of Springfield, Vermont* (Englewood Cliffs, N.J., 1959); and Harless D. Wagoner, *The U. S. Machine Tool Industry from 1900-1950* (Cambridge, 1966).

11. Robert S. Woodbury, *History of the Gear Cutting Machine* (Cambridge, Mass., 1958), *History of the Grinding Machine* (Cambridge, Mass., 1959); *History of the Milling Machine* (Cambridge, Mass., 1960); *History of the Lathe* (Cambridge, Mass., 1961).

12. Edwin A. Battison, "Eli Whitney and the Milling Machine," *The Smithsonian Journal of History* 1 (Summer 1966): 9, 16-23; and "A New Look at the 'Whitney' Milling Machine," *Technology and Culture* 14 (1973): 592-98; Felicia Deyrup, *Arms Makers of the Connecticut Valley* (Northampton, Mass., 1948); Merritt Roe Smith, "John H. Hall, Simeon North and the Milling Machine: The Nature of Innovation among Antebellum Arms Makers," *Technology and Culture* 14 (1973): 573-91; and Paul Uselding, "Henry Burden and the Question of Anglo-American Technological Transfer in the Nineteenth Century," *Journal of Economic History* 30 (1970): 312-37, and "Elisha K. Root, Forging, and the 'American System,'" *Technology and Culture* 15 (1974): 543-68.

13. Merritt Roe Smith, *Harpers Ferry Armory and the New Technology* (Ithaca, N.Y., 1977), chaps. 2, 3, and 4.

14. Included here is a representative listing of some of the better-known *American Machinist* articles: E.G. Parkhurst, "One of the Earliest Milling Machines," 23 (1900): 217-25; E.A. Dixie, "Some More Antique Machine Tools," 31 (1908): 558-60; Luther D. Burlingame, "The Development of Interchangeable Manufacture," 47 (1914): 295-97; and Guy Hubbard, "Development of Machine Tools in New England" (various articles), 59 (1923), 60

(1924), 61 (1924). Other periodicals that carried articles on machines and machine practice in this period are *Iron Trade Review, Machinery, Engineer, Iron Age, Scientific American,* and *Proceedings, Institution of Mechanical Engineers.*

15. H.J. Habakkuk, *American and British Technology in the Nineteenth Century* (Cambridge, 1962); and David Landes, *The Unbound Prometheus* (Cambridge, 1969).

16. Paul Uselding, "Studies of Technology in Economic History," in *Recent Development in the Study of Business and Economic History: Essays in Memory of Herman E. Krooss,* ed. Robert E. Gallman, published as suppl. no. 1 to *Research in Economic History* (Greenwich, Conn., 1977), pp. 159-220.

17. Charles Babbage, *On the Economy of Machinery and Manufactures* (London, 1832).

18. James Nasmyth, *Autobiography,* ed. Samuel Smiles (London, 1883), p. 146.

19. Roe, *English and American Tool Builders* (note 8), p. 99.

20. Ibid., p. 101. Also see John Fernie, "Decimal System of Measurement with Description of Mr. Whitworth's Measuring Machine and Standard Gauges," *Proceedings of the Institution of Mechanical Engineers,* 1859, pp. 110-22.

21. Ambrose Webster, "Early American Steel Rules," *American Machinist* 17 (1894): 7.

22. Quoted from ibid.

23. Ibid.

24. E.A. Dixie, "Some Old Gages and Filing Jigs," *American Machinist* 38 (1908): 381-83. By the 1840s the Springfield Armory kept three sets of gauges: one set used by workmen received the most wear and therefore was most likely to "go out of gauge"; the second set was used by inspectors to check pieces turned out by the workmen; finally a master set was retained to check the inspector's work and to serve as "the court of last appeal."

25. Ibid., p. 382.

26. Lucian Sharpe, "Letter of December 10, 1887," *American Machinist,* printed as "Development of the Micrometer Caliper," 15 Dec. 1892, pp. 9-10. For a similar account see *The Micrometer's Story, 1867-1902* (Providence, n.d.).

27. Notwithstanding Luther Burlingame's report that a Brown and Sharpe steel scale from the mid-1850s was found, in 1923, to have no greater than 0.0001–0.0002-inch variation from standard over its 24-inch length. "Pioneer Steps Toward the Attainment of Accuracy," *American Machinist* 47 (1914): 237-43.

28. John E. Sweet, *Machine Practice* (Ithaca, N.Y., 1875).

29. Frederick A. Halsey, *Methods of Machine Shop Work* (New York, 1914), p. 84.

30. Burlingame, "Pioneer Steps Toward the Attainment of Accuracy" (note 27), p. 241.

31. George M. Bond, "Standards of Length Applied to Gauge Dimensions," *Journal of the Franklin Institute* 117 (1884): 379.

32. Torsten K. W. Althin, *C. E. Johansson 1864-1943, The Master of Measurement* (Stockholm, 1948), p. 105.

33. Wayne R. Moore, *Foundations of Mechanical Accuracy* (Bridgeport, Conn., 1970), pp. 156-57.

34. Ibid., p. 157.

❋❋❋❋❋❋❋❋❋❋❋❋❋❋❋

THE *SYSTEM:* THEORY AND PRACTICE

David A. Hounshell

Despite a substantial body of scholarship, often first-rate, on the American system of manufactures, historians have failed to understand and to define this phenomenon adequately. The major reason is that the American system of manufactures has been identified too exclusively with antebellum small-arms manufacture by and for the federal government.

Certainly there is much to justify the equation of small-arms production methods with "the American system of manufactures." As Merritt Roe Smith has demonstrated in his paper for this symposium and in his recent book on Harpers Ferry, government sponsorship played a decisive role in the creation of what was in 1850 a high technology of small-arms production.[1] Two principal characteristics distinguished this production technology. First, there was a high degree of mechanization, specifically the use of special-purpose or single-purpose machine tools operated sequentially to carry out a series of manufacturing operations. Second, small-arms production by and for the federal government was marked by a quest for interchangeability of parts. This quest resulted in the establishment of important principles of and procedures for precision manufacture, specifically a rational jig and fixture system and a rigid system of gauging. Precision manufacture manifested itself at armories in bureaucratized structures through methods which set small-arms production apart from other areas of manufacture in the United States in the years before 1850—manufactures such as axes, clocks, barrels, furniture, and farm implements.

These areas of American manufacture were often characterized by the same high degree of mechanization as arms production—by the use of special-purpose machines—but nowhere are to be found the elaborate, bureaucratized elements of precision manufacture practiced in the nation's armories. Are we to say, therefore, that axemaking or furniture manufacture was not carried out under the American system of manufactures? Certainly not. I am suggesting that if historians define the

American system of manufactures on the exclusive basis of small-arms production, they fail to understand what the British observers Joseph Whitworth and John Anderson meant by the expression. What struck them as most characteristic of American production technology was, in Anderson's words, "the adaptation of special tools to minute purposes."[2] For Whitworth, interchangeability of parts remained almost completely inconsequential even when he was reporting on the Springfield Armory.[3] Anderson studied the question of interchangeability more carefully— perhaps only because he was so ordered—but strictly within the context of the national armory at Springfield and other small-arms factories. For Anderson, the manufacturing practices at these armories constituted a very special instance of the American system of manufactures, not the general one, the key to which was mechanization rather than interchangeability. Interchangeability of parts, with its requisite principles and practices of precision manufacture, constitutes something that is most appropriately called "New England armory practice."[4]

Until historians are willing to make this distinction between the broader expression "the American system of manufactures" and the more specific one "armory practice," the former will remain an imprecise and ambiguous phrase at best. And if historians insist upon equating New England armory practice with the American system of manufactures, they will misinterpret the technologies employed in the New England wood- and brass-clock industry; in axemaking at Collinsville, Connecticut; in furniture manufacture in Cincinnati; and in pin manufacture and a host of other areas of highly mechanized production.

I am not suggesting that the study of New England armory practice be abandoned, for I am deeply committed to its study. Rather, I am simply pleading for a recognition that the manufacturing technology developed in New England armories is a quite special manifestation of the American system of manufactures.

What makes armory practice so special? As I have already suggested, the principles and practices of precision manufacture—of the quest for interchangeability through the development of a rational jig and fixture system and a rigid system of gauging—make the difference. By rational jig and fixture system, I mean the following: A fixture is a device which "fixes" or secures a workpiece in a machine tool, such as a milling machine, and holds it during the machining operation. Each time a workpiece is fixed in a machine tool, a certain amount of inaccuracy creeps in. With the American system of manufactures and its characteristic of numerous machining operations on a single part—each operation carried out in a different machine tool and usually in a different fixture—a multiplying effect sets in; the inaccuracy of each fixing is multiplied by the number of different machining operations on a given part. John Hall thought of a way to nullify this multiplying effect by locating the workpiece in each fixture relative to a single reference point on the piece. Hall called this the bearing or location point, and his principle was adopted almost universally.[5] Jigs and fixtures designed with reference to a single bearing point constitute therefore a rational jig and fixture system.

Gauges which check or verify dimensions of parts must also be designed with reference to bearing points used in fixture design. At the national armories, gauging had become a bureaucratized system by the 1850s. The entire system was based on a model or pattern weapon. I do not use the word model in the same way I do when I say I drive a 1972 Opel Model 1900. For armsmakers, *model* was used in the Platonic sense of an *ideal form*. Production arms were imperfect imitations; the gauge system was an attempt to make them more like the model. Three sets of gauges conforming to the model were constructed to check critical dimensions of the production arms. These gauge sets were called the master gauges, the inspection gauges, and the production gauges. Master gauges were used to check the inspection gauges, which in turn verified the production gauges. Since gauges (especially production gauges) were subject to wear through daily use, elaborate inspection procedures—one might even call them rituals—evolved at the armories so that precision might be maintained. Precision manufacture grew increasingly bureaucratized in the late antebellum period, and, after the Civil War, when large factories arose which operated on the armory system even though they did not make firearms, many established separate tool departments and gauge departments. In some cases, gauge maintenance constituted a separate department too.[6]

These specific techniques of precision manufacture clearly distinguish armory practice. In the antebellum period they are unique to the nation's armories. During the second half of the nineteenth century, when a mechanic said that a factory—whether a sewing-machine plant, a bicycle works, or an agricultural-implement factory—operated with "armory practice," he meant a system of manufacture employing a rational jig and fixture system, and a model-based gauge system.

With this distinction in mind I would like to examine the introduction of armory practice into other areas of manufacture beginning about 1850. Contemporary "system thinking" is perhaps best seen in such instances. Specifically, I wish to consider armory practice and the nascent sewing-machine industry. This is appropriate for a number of reasons, the most significant being that the sewing-machine industry was the first to adopt armory techniques of production. Secondly, there has been a good deal of attention paid to the sewing machine recently, much of it tending to perpetuate old misunderstandings. This tendency is epitomized in Ruth Brandon's popular biography of Isaac Singer published in 1977.[7] A *Newsweek* review distilled Ms. Brandon's interpretation of Singer, the American system of manufactures, and interchangeable parts: "[Isaac Singer] pioneered mass production of products with interchangeable parts by inventing machines to manufacture machines."[8] Even Paul Uselding recently lent credence to this misapprehension: "The story of American prowess at the Crystal Palace, the displays of the Colt Revolver, the McCormick Reaper, the Singer Sewing Machine—all embodying the principle of interchangeability—is now a familiar tale.'"[9] A familiar tale, to be sure, but not necessarily true. There is much to be gained by a careful study of the sewing-machine industry and its production technology from the 1850s to the 1880s.[10]

I have studied the history of three sewing-machine companies which were formed in the 1850s and which continued to operate into the twentieth century. Two of the companies, Singer and Willcox & Gibbs, survive today, while a third, Wheeler & Wilson, was absorbed by Singer in 1907. The early history of these companies is essentially a study in technical choice and contemporary "system thinking."

In the early 1850s manufacturer/capitalist Nathaniel Wheeler teamed up with inventor Allen B. Wilson to form the Wheeler and Wilson Manufacturing Company with a capital of $160,000. Though this firm led the sewing-machine industry in sales until 1867 (when Singer made a runaway move toward complete dominance), it got off to a slow start. By mid-1853 it had manufactured no more than 300 machines, each unique, built one by one with a small line of general metalworking tools in a job shop. The company might have continued to produce sewing machines in this fashion had Wheeler not hired William Perry as superintendent. Wheeler himself had remained largely outside the sphere of development of New England manufacturing technology and certainly outside the development of armory practice. But Perry had learned the machinist's trade at Samuel Colt's Hartford Armory and had become one of Colt's contractors.[11] Colt, it may be recalled, had built one of the best-known showplaces of American production technology. His armory provided a training ground for men who spread armory practice throughout the nation in the second half of the nineteenth century.[12]

In introducing armory practice at Wheeler & Wilson's sewing-machine factory, William Perry relied not only on his own know-how but also on that of two other armsmakers, Joseph Dana Alvord and James Wilson (no relation to Allen B.).[13] Alvord had learned the machinist's trade under Nathan P. Ames of Chicopee, Massachusetts, and had then worked at the Springfield Armory for eight years. In 1851 he had moved from Springfield to Hartford, where he worked as a contractor at the Robbins & Lawrence-Sharps rifle factory. That year, Robbins & Lawrence, a firm based in Windsor, Vermont, had gained an international reputation by its display of firearms with interchangeable parts at the London Crystal Palace Exhibition.[14] Perry also hired James Wilson, Alvord's friend and a fellow Robbins & Lawrence contractor. Together, in 1857, Perry, Alvord, and Wilson set up for Wheeler & Wilson a sewing-machine factory in Bridgeport modeled entirely after amory practice, especially the Colt version in which major components were initially drop-forged. By 1862 these armory mechanics had created a plant capable of producing almost 30,000 sewing machines annually. The company claimed that with its accurate, specialized machine tools, its rational jig and fixture system, and its extensive gauges, parts of the Wheeler & Wilson sewing machine never needed a "stroke of a file" during assembly.[15] From 1857 until 1872, when production peaked at 174,088 machines, Wheeler & Wilson developed an increasingly refined version of New England armory practice.[16] One might even call the Wheeler & Wilson system "high" armory practice.

While the Wheeler & Wilson sewing machine had initially been the product of skilled machinists in a job shop and then become a uniform article turned out by a factory operating under the armory system of manufacture, the Willcox & Gibbs sewing machine was never manufactured any other way except under armory practice. Founded in 1857, the Willcox & Gibbs Sewing Machine Company represented the partnership of inventor James E. A. Gibbs and capitalist James A. Willcox.[17]

In March 1858 Willcox asked J. R. Brown and Lucian Sharpe—who ran a small job shop in Providence—to make twelve Willcox & Gibbs sewing machines, promising them more business if the machines were "got up right" and if they succeeded in the market.[18] With its three small engine lathes, two hand lathes, an upright drill, a hand lever planer, and a "donkey" planer,[19] one would have expected Brown & Sharpe to build these twelve sewing machines one by one. This was not the case, however. Before they ever completed a commercial sewing machine, Brown & Sharpe set about designing a model, along with special tools, jigs, fixtures, and gauges. The reason is not readily apparent since neither partner had had any experience with the American system of manufactures in general or armory practice in particular. Only in a letter to Willcox in which Sharpe sought to justify the delay in completing sewing machines is there the slightest hint as to the reason Brown & Sharpe chose to pursue armory practice. Sharpe wrote: "Things appear to go slow it is true . . . [However], one of our best men who formerly worked at the Sharp's [*sic*] Rifle Company in Hartford says that we have got along with the tools as fast as he expected. He has seen $10,000 spent upon tools for a single gun lock and $25,000 for tools for a rifle."[20] The workman who had had firsthand experience with armory practice must have persuaded Brown & Sharpe that the techniques used to produce Sharps rifles could be used to manufacture Willcox & Gibbs sewing machines.

While this was true in theory, Brown & Sharpe found that actual implementation of armory practice was not simple. They were amazed at the length of time it took to build the necessary hardware, and its costliness.[21] Nonetheless, they kept faith, as is evident from a letter Lucian Sharpe sent James Willcox six months after undertaking to make sewing machines:

> We do not wonder at your surprise that the machines are not in a greater state of forwardness, yet if we had turned out a quantity in an imperfect state as would have been the case had they been made without the templets and other tools it would without doubt [have] injured the reputation of the machine if it did not kill it entirely. Our first trouble was in getting a hook to work and after that was accomplished we made tools by which we could turn out any quantity of machines.... Tools for all the operations have been made with the same regard to producing perfect work as in the templets for drilling, and the time and cost have ten times exceeded our expectations. By the experience now acquired we now hope to turn out nearly perfect machines at the outset instead of proceeding in the usual way with such things.[22]

Eventually Brown & Sharpe succeeded, and the firm continued to manufacture sewing machines for the Willcox & Gibbs Sewing Machine Company until the 1950s. In the nineteenth century, production reached a peak of 33,639 machines, in 1872, the same year Wheeler & Wilson peaked at 174,088. In 1872 the Singer Manufacturing Company produced 219,758 sewing machines, but that was far from a peak. By 1880 it was making half a million.[23] Yet Singer's production technology in its first twenty years scarcely evidenced a potential for achieving such a prodigious output.

The origin of the Singer Manufacturing Company is, once again, rooted in the partnership of an inventor and a capitalist, Isaac Merritt Singer and Edward Clark.[24] Originally the Singer machine had been made in Boston by Orson C. Phelps, proprietor of a philosophical instrument shop.[25] For a few days after Singer had built his first machine—in Phelps's shop— Phelps had held an interest in the company hastily formed to exploit the invention. Phelps soon sold out to Singer, but continued to make his sewing machines. Because he simply was not equipped to produce cast-iron machines weighing over 125 pounds, Phelps obtained major components from a local founder, jobbed out lathe and planing work, purchased necessary bolts and nuts from hardware merchants, and fitted everything together in his shop with files and grindstones.[26]

Singer's New York factory in 1854. Note the paucity of machine tools compared to the large number of hand filers and fitters. (From United States Magazine, *15 Sept. 1854.)*

This piecemeal approach continued even after Singer teamed up with Clark in 1852 and after they moved operations to a workshop in New York City. Castings, spur gears, and nuts and bolts were obtained from local shops and merchants. Seeking to do some of its own machining work, I.M Singer & Company purchased a few small lathes, planers, and boring machines. What dominated the company's tool expenditures, however, were vast quantities of British-made files and files recut by a New Jersey firm.[27]

An illustration of the Singer factory, printed in 1854 in *United States Magazine* (above), shows some power tools in operation, but more importantly it depicts dozens of workmen standing at benches filing away at parts held in vises.[28] Nowhere—not in this illustration, or in published materials, or in company manuscripts—do we see or hear about the special-purpose machine tools of the American system of manufactures or the modes of attaining precision characteristic of armory practice. The Singer machine was "produced by hand at the bench."[29]

While other manufacturers such as Colt, Wheeler & Wilson, and Brown & Sharpe attributed their success to the way their products were made (i.e., armory practice), I.M. Singer & Company emphasized a different ingredient. As the company's leader, Edward Clark, put it: "A large part of our own success we attribute to our numerous advertisements and publications. To insure success only two things are required: 1st to have the best machines and 2nd to let the public know it." [30] "To have the best machines" implies not only excellence in design but also quality in manufacture. There was no question in Clark's mind that the Singer method of manufacture—a method very different from the American system of manufactures—provided sufficient quality for commercial success.

I.M. Singer & Company's choice of production methods was dictated largely by Isaac Singer's and Edward Clark's ignorance of the American system of manufactures. There is little in their background which would indicate contact with either that system or with armory practice.[31] Moreover, Clark and Singer never hired a mechanic who had worked in a New England armory or a factory operating under the American system of manufactures. Ignorance of the system is indicated by an episode which occurred in 1855. That year I.M. Singer & Company set up what proved

Singer advertisement, 1857. (Courtesy of Eleutherian Mills Historical Library.)

Singer showroom, 1857. (Smithsonian Institution Neg. No. 48091–B.)

Singer sewing machine, serial number 22 (1851), submitted as a patent model. (From the collections of the Division of Textiles, National Museum of American History; Smithsonian Institution Neg. No. 45572–D.)

to be an abortive factory in Paris, France, under the direction of William F. Proctor, a workman from the New York factory. In a letter to Proctor, Clark listed the machine tools the company planned to send to Paris—a few lathes, hand planers, and boring machines. Clark's list could have been an inventory of machine tools in any job shop in America or Europe, yet he believed he was fitting out "an establishment which shall be a model one, and a credit to American skill in the mechanical arts."[32] Perhaps Clark partook of the sort of chauvinism exhibited by Samuel Colt and Alfred C. Hobbs; at any rate, he was enjoying vicariously the praise heaped upon these men by the British press after the London Crystal Palace Exhibition.[33] Still, there is simply no comparison to be made between Colt's model armory established in London in 1853 and the envisioned "model" sewing-machine factory Singer set up in Paris. With Colt we see a New World factory in an Old World setting; with Singer, an Old World factory in an Old World setting.

I.M. Singer & Company built its own factory in New York in 1858. But unlike the Wheeler & Wilson Sewing Machine Company when it moved into a new factory at Bridgeport in 1857, Singer adopted neither the American system of manufactures nor the more specialized armory practice. For the next fifteen years, at least, as B.F. Spalding pointed out long ago in the *American Machinist*, Singer "compromised with the European method [of manufacture] by employing many cheap workmen in finishing pieces by dubious hand work which could have been more economically made by the absolutely certain processes of machinery."[34]

Not until 1862, when the Singer company hired a Yankee mechanic named Jerome Carter to make its needles, did the factory use *any* special-purpose machinery.[35] The year after this initial venture with the American system of manufactures, the Singer officials moved on to a first encounter with armory practice when they hired Lebbeus B. Miller to "design and supervise the construction of special tools for the production of interchangeable parts."[36] In 1863 Singer made slightly over 20,000 sewing machines, and perhaps had reached its limit in terms of output with hand methods. Or perhaps costs had become too great. In any case, Miller had been exposed to a New Jersey version of armory practice, having worked for the Manhattan Firearms Company while Andrew R. Arnold was the factory superintendent. Arnold had come to Manhattan from Colt's Hartford Armory. Miller, who had been an apprentice of Ezra Gould, a New Jersey machine-tool maker, worked for Arnold at Manhattan for two years before taking the job with Singer. It is almost certain that he never saw at Manhattan a full-blown version of New England armory practice such as at Colt's or at Springfield.[37] He was at Singer for twenty years before he instituted full armory practice. In the first decade his most significant contributions were the introduction of drop-forging and the construction of special-purpose machine tools. Not until after 1873 does Miller appear to have introduced a model-based gauge system comparable to that of the better armories.[38]

With the Singer company I believe we have a notable example of the absence of "system thinking." For over a decade Singer made its machines completely by European methods, with general-purpose machine tools and a great amount of hand labor. Unlike many sewing-machine companies, it survived. Singer directors placed great emphasis on marketing, but more importantly they believed that hand methods produced the most satisfactory sewing machine. From 1863 to 1873, the company adopted piecemeal the American system of manufactures, relying increasingly on special-purpose machine tools. This came about only after a Yankee mechanic and a local machinist were hired who knew a little about how things were done at the Colt Armory. Between 1868 and 1870, when the Singer directors were discussing construction of a new factory, George Ross McKenzie, the company's vice-president, suggested taking a look at operations of the Wheeler & Wilson factory and the Springfield Armory.[39] But not until after 1873, when this new factory opened, did significant efforts to attain precision manufacture take place at the Singer Manufacturing Company. By that time the company was making over 230,000 sewing machines annually and had begun to realize that without greater uniformity of parts further multiplication of output would entail massive problems.

By 1880 all three sewing-machine companies displayed a high degree of system thinking. The quantity of output varied tremendously between the three, and output obviously influenced the way in which work was organized and the way material flowed through the factories. While Singer's Elizabethport works turned out over a quarter of a million sewing machines, Brown & Sharpe made less than 20,000 Willcox & Gibbs machines. Wheeler & Wilson fell midway between these extremes.[40] The very notion of using special-purpose machine tools sequentially suggests a pattern for the flow of work through a factory. For Brown & Sharpe, which in its peak year made less than 34,000 machines, or about 126 per day if we count 45 six-day weeks, this natural sequence could not be followed. Let me illustrate. A sketch of Brown & Sharpe's sewing-machine operation (p. 138, top) was made by George E. Whitehead, who worked for the company from 1865 to 1868.[41] As is evident from Whitehead's sketch, Brown & Sharpe had a limited number of machine tools, at least in the 1860s, for making the Willcox & Gibbs machine. Obviously, the machine tools had to be set up more than once to finish parts which required ten or twenty operations.

Henry Leland, a master of precision machine work who later created the Cadillac Motor Car Company, headed Brown & Sharpe's sewing-machine department from 1878 to 1890. Previously, Leland had worked as a tool builder at the Springfield Armory and at Colt's Hartford Armory. While at Brown & Sharpe, he introduced operation sheets—lists which enumerated all machining operations on parts and the machine tools employed, as well as the necessary tools, jigs, fixtures, and gauges.[42] Operation sheets (which may date back to the 1830s) are indicative both of system thinking and a concern for the organization and flow of work. A *Scientific American* illustration of sewing-machine production at Brown & Sharpe's new factory in 1879 (p. 139, top) shows it still isolated in a single room, away from the company's other manufacturing operations.[43]

Wheeler & Wilson used the same single-room approach to production, as is evident from another *Scientific American* engraving, likewise from 1879 (p. 139).[44] Although detail is sketchy, the factory illustration shows quite clearly how much of the work process was determined simply by the layout of the lineshafting system.

As for Singer's techniques of manufacture, we have a very early graphic and several from 1880. The representation of the factory in 1854 (p. 133), shows relatively few machine tools compared to the army of workmen at the bench with files in hand. I have found no other published illustrations of Singer factory operations until more than a quarter of a century later (and I have often wondered whether officials were not deliberately responsible for this lacuna). In 1880, however, an excellent series of illustra-

Floorplan of Brown and Sharpe's sewing-machine room (ca. 1865), after an original sketch by a former employee. Note the limited number of machine tools used in producing the Willcox and Gibbs sewing machine. (Courtesy of Brown and Sharpe Manufacturing Co.)

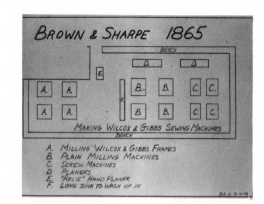

Brown and Sharpe factory in Providence (1860s), with the workers on the upper floor displaying Willcox and Gibbs machines. (Courtesy of Brown and Sharpe Manufacturing Co.)

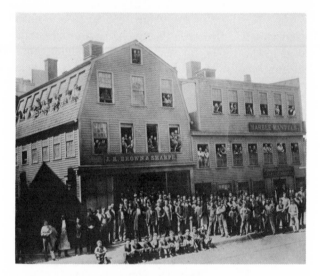

Willcox and Gibbs sewing machine. (From the collections of the Division of Textiles, National Museum of American History; Smithsonian Institution Neg. No. P6393.)

Sewing-machine room
of the Brown and
Sharpe factory, 1879.
(From Scientific
American, 1 Nov.
1879.)

Wheeler and Wilson
factory, 1879. (From
Scientific American, 3
May 1879.)

Wheeler and Wilson
sewing machine, ca.
1876. (From the col-
lections of the Division
of Textiles, National
Museum of American
History; Smithsonian
Institution Neg. No.
17663–C.)

tions appeared in a Singer-supported book, *Genius Rewarded; or the Story of the Sewing Machine.* [45] These engravings show quite clearly the transformation that had occurred in Singer manufacturing technology. Whereas most parts of the Singer machine had formerly been cast iron, L.B. Miller relied increasingly on drop forging and fabrication from bar stock (see below). The major exceptions were the base and arm, cast in the foundry shown opposite. After Miller became superintendent, Singer acquired ever-larger numbers of special-purpose machine tools, especially

Forging shop, Singer factory, 1880. (From Genius Rewarded; or the Story of the Sewing Machine *[1880].)*

Screw-manufacturing department, Singer factory, 1880. (From Genius Rewarded; or the Story of the Sewing Machine *[1880].)*

Foundry, Singer factory, 1880. (From Genius Rewarded; or the Story of the Sewing Machine *[1880].)*

milling machines and screw machines. Miller reported in 1885 that "by the addition of machinery the same number of employees will produce double the number of mchs. [machines] they would ten years ago." [46]

Symbolic of the greater precision obtained in production, sometime between 1874 and 1880 Singer changed the name of the department in which the sewing machines were put together from the Fitting Department to the Assembling Department.[47] In its public-relations literature after 1880, the company began bracketing the word "assembled" in quotation marks to describe the process by which the machines were put together.[48] As late as 1885, however, Singer called the employees who worked in the assembling department "fitters" and paid them, on average, 40¢ more per day than ordinary machine tenders.[49] Literature of a later date, the 1890s, would play up "The Singer Assembly System." [50]

Finally, something needs to be said about the American system of manufactures and scientific management. If by scientific management we mean Taylorism, then there is nothing to say. Taylorism is a historical phenomenon largely divorced from the development and diffusion of the American system of manufactures in general and armory practice in particular.[51]

Assembly room, Singer factory, 1880. (From Genius Rewarded; or the Story of the Sewing Machine *[1880].)*

Recall Henry Ford's indictment of scientific managers in his *Encyclopaedia Britannica* article on mass production.[52] The pioneers of the American system of manufactures probably would have shared Ford's attitude. But if by scientific management we mean an approach to factory management peculiar to the American system of manufactures, then I think that this is getting at the very heart of the issue.

I would like to end by recounting briefly an episode in the Singer Manufacturing Company's production history which demonstrates the critical role played by management. The episode occurred in the early 1880s when, as I have said, the company manufactured sewing machines with its peculiar version of armory practice. Singer was then trying to put into production a new, more sophisticated model of sewing machine—the Improved Family—while simultaneously increasing its output of existing models and also supplying the company's Scottish factory with additional machine tools. The Elizabethport factory was supposed to turn out eight thousand machines per week. Yet seldom was this number attained, despite the best intentions of workers and executives.[53]

After several months of patient waiting for the initial production of the new model machine, Singer's president, Edward Clark, and the vice-president, George Ross McKenzie, decided that Lebbeus Miller, the superintendent at Elizabethport, could not handle all the details of factory operations along with initiating production of the new model. To take some of the burden off Miller, McKenzie hired Albert D. Pentz to oversee production of the new model and particularly to determine inspection procedures for this machine. Pentz had worked at the Singer factory in the early 1870s before taking charge of Singer's extensive repair shop in Chicago. There he had invented several sewing mechanisms, some of which were incorporated into the new model. When Pentz came to the Elizabethport works in March 1881, McKenzie asked him to convey his general "impressions of the status in the factory"[54] and to suggest how its operations might be improved. In today's parlance, Pentz was to be Singer's first industrial engineer.

Complying at once with McKenzie's request, Pentz identified Singer's inside contracting system as the root of both success and failure.[55] Thus far I have not mentioned the inside contracting system, but Alfred Chandler's work and that of Daniel Nelson necessitate consideration of this early form of factory management.[56] In 1952 John Buttrick published an article on "The Inside Contract System" in the *Journal of Economic History*, and since that time this feature of the American system of manufactures has received little fresh attention.[57] Buttrick suggested that the origin of

First introduced in 1865, Singer's New Family Sewing Machine became the company's major seller until the 1880s. (From the collections of the Division of Textiles, National Museum of American History; Smithsonian Institution Neg. No. 58987.)

this system is rooted with "the increasing use of the principle of inter-changeable parts,"[58] though I have seen inside contracting more wide-spread than this. Nonetheless, since Eli Whitney, Robbins & Lawrence, Brown & Sharpe, Colt, Remington, Winchester, Singer, Wheeler & Wilson, and other shops employing armory practice used this system, it is fair to say that inside contracting provides another aspect of armory practice.

Singer used a relatively small number of inside contractors. Hans Reiss was the contractor for both the machining and the assembly departments. William Smith held the contract for inspecting parts and finished machines, while William Inslee was the contractor for the adjustment of the assembled sewing machines. Of course, other men contracted for more specialized processes such as shuttle production, japanning (i.e., painting), needlemaking, and crating. When Albert Pentz assumed responsibility for getting the new Singer model into production, his principal task was to coordinate the activities of the various contractors and, more importantly, to determine limits of precision and standards for inspection.

His trials and tribulations were many. He found the contractors at Elizabethport belligerent and hard to deal with, but, worse, he detected weaknesses in the very management of the factory. The following letters from Pentz to McKenzie perhaps best give an idea of management troubles at Singer. The first was dated June 13, 1881:

Immediately prior to my coming here, one of the workmen employed on the assembly floor, in Mr. Reiss' department, for some unknown reason scraped the tables [or bases] of the I.F. [Improved Family, the new model sewing machine] where the throat plate sets in on the edge to such an extent that there was $\frac{1}{32}$ of an inch too much room for the throat plate. These machines began to come back from the japanning [kilns] about two weeks ago. When I discovered these I condemned them all. Whereupon Mr. Reiss appealed to Mr. Diehl [one of the super-intendent's assistants] who said they could all go and ought to be accepted. I then took the case to Mr. Miller [the superintendent], who said they should all be thrown out. But in the meantime, I, remembering your advice to me, to not be too particular with points which had no practical value, and the only thing this affects is the looks of the machine, the screws determining the position—proposed as a compromise that we accept all which did not pass the limit of what I thought would not hurt the appearance of the machine. Mr. Miller agreeing to this gave Mr. Inslee's man, who examines the machines that are received from Mr. Reiss, a model machine for a limit beyond which we would not accept from him. While I was away last week it was found convenient to loose [sic] that model and to put through a large number of machines which are too much open for acceptance. Mr. Smith at this juncture discovered the thing and says that *none* of the machines are good enough to be accepted. The inspectors say that they get as [so] many different kinds of instructions that they don't know what to do.[59]

A second letter from Pentz was dated near the end of June 1881:

When I came here I examined the I.F. and fixed a temporary standard of inspection and instructed the inspector to that standard. A short time afterward, I discovered that the races were coming excentric [sic] to the shuttle shaft, and a number of other points were wrong and of course I went for the inspector who said that Mr. Smith had changed the standard that I had given him. This Mr. Smith denied and I was inclined to believe him until the following series of annoyances occurred.[60]

Pentz went on to describe in detail this series of "annoyances," which make it obvious that no one operated under firmly established lines of authority. Earlier he had reported that "no one appears to know just how perfect [Singer machines] should be."[61]

About this same time, one of McKenzie's assistants had informed him that "it is perfectly clear to me that before there is to be a new machine, you will have to go and live down at the factory for a few months."[62] Seven weeks later this same junior executive reported that, while hundreds were expected, the factory had turned out only two of the new machines in a whole week due to the lack of two parts and "to Pentz's inspection being unusually severe."[63] With Singer's inside contracting system, continual problems arose with inventory control of parts as well as with inspection of those parts.

For almost two more years the Singer factory operated with the same chaotic system of factory management. Finally, in 1883 George Ross McKenzie, now president of the company, began to sort out factory management problems. First of all, he hired Edwin H. Bennett to serve as assistant superintendent to L.B. Miller.[64] Working together as a committee, McKenzie, Miller, Bennett, Pentz, and Philip Diehl, another Miller assistant, attacked some of the deep-seated problems of the Singer production system. Unquestionably the most important change these gentlemen made is revealed in the following resolution they adopted in March 1883:

Decided that each piece commenced in a department shall be finished there to guage [sic] ready for assembling and no part shall be made in the department where it is assembled into the machine.[65]

It is important to note that this is an *administrative* or *managerial* decision, not a purely technical one.[66]

While the efforts of this high-powered manufacturing committee began to show results, they were not entirely unambiguous. Late in 1884 Pentz remarked that the factory was "no further ahead than we were two years ago" in determining standards of quality.[67] However, Edwin Bennett identified the real problem of the Singer factory:

> I am more convinced than ever [he wrote McKenzie] that the best results from this [Improved Family] machine can only be obtained by close workmanship and the application of tools with the least possible amount of hand labor in their production. To accomplish this result, I believe, will require the services of a thorough practacle [sic] mechanic to superintend the inspection of all operations, parts, and gauges. This position would be very similar to the one now held by Mr. Pentz, only we want less science and more practice. No changes of model or gauge points should be allowed except in committee of the whole, where a complete record should be kept, all changes [recorded] and reason for such.[68]

Bennett himself assumed the role of that "thorough practical mechanic." During the next two years he wrote a detailed bluebook for the Elizabethport factory which delineated all machining operations and work-flow routes for the Improved Family sewing machine. Going beyond simple operation sheets such as those used by Henry Leland at Brown & Sharpe, Bennett's bluebook also specified inspection procedures and limits of precision. In sum, this bluebook codified Singer factory managerial policy for the first time in the company's history, and it also made clear— implicitly, at least—the line and staff arrangements of the factory.[69] It is of some importance, I think, although perhaps not causally connected, that the emergence of the Singer bluebook marks the end of the inside contracting system at Singer's Elizabethport factory. Rate disputes seem to have been the immediate cause for the abandonment of inside contracting, but it seems to me that the underlying cause was that series of problems inherent in this system, problems enumerated if not elaborated by John Buttrick.[70]

In this essay I have tried to show the necessity for making a distinction between the broader concept of the American system of manufactures and its special case, armory practice. Moreover I have sought to illustrate the way in which three different sewing-machine manufacturers chose their respective production technologies. It is clear from these case histories and others in wideranging fields of manufacture that the mechanician who had worked in an armory or had worked in a factory using armory practice played a critical role in making this practice applicable in new areas of production—in the manufacture of different consumer durables.[71] Finally, the case of the Singer factory—one of the most celebrated factories of the nineteenth century—illustrates how "system thinking" is revealed in factory management. The operation sheets used by Brown & Sharpe and by Singer suggest one level of system thinking. Yet the very nature of the inside contracting system, if Singer is a good example, showed a marked tendency of *disintegrated* system thinking. With Singer we see movement from the by-no-means-simple meeting of the technical demands of armory practice and the elaboration of a specific set of rules and policies designed to determine the organization of work and to facilitate the flow of that work. While perhaps not critical for the success of armory practice (witness Singer's vast production in the early 1880s despite the chaos in its factory managerial procedures), the development of mass production in the twentieth century would rest firmly on the codification of system thinking as reflected in Edwin Howard Bennett's bluebook of the Singer Improved Family sewing machine.

Notes

1. *Harpers Ferry Armory and the New Technology* (Ithaca, N.Y., 1977).

2. "Report of the Committee on the Machinery of the United States of America," reprinted in Nathan Rosenberg, *The American System of Manufactures* (Edinburgh, 1969), p. 193.

3. "Special Report of Mr. Joseph Whitworth," reprinted in *The American System of Manufactures* (note 2), pp. 327–89. Whitworth mentions interchangeability only briefly in his section on the Springfield Armory. He wrote (p. 365): "The complete musket is made (by putting together the separate parts) in three (3) minutes. All these parts are so exactly alike that any single part will, in its place, fit any musket."

4. This concept, which I first encountered in numerous articles in *American Machinist* from the 1890s, is more fully developed in my doctoral dissertation, "From the American System to Mass Production: The Development of Manufacturing Technology in the United States, 1850–1920" (University of Delaware, 1978).

5. Hall's thinking about location points is admirably discussed in Smith's *Harpers Ferry Armory and the New Technology* (note 1), pp. 226–27.

6. The history of gauging has not been written, but Paul Uselding has begun to deal with the subject in this volume. Smith, *Harpers Ferry Armory and the New Technology* (note 1), pp. 109–10, 225–26, 231, 280, 284 deals with gauging in antebellum arms manufacture perhaps more extensively than any other source. See also Charles H. Fitch, "The Rise of a Mechanical Ideal," *Magazine of American History* 11 (1884): 516–27, and Fitch's "Report on the

Manufactures of Interchangeable Mechanism," in *Report on the Manufactures of the United States at the Tenth Census* (Washington, D.C., 1883), p. 3.

7. *A Capitalist Romance: Singer and the Sewing Machine* (Philadelphia, 1977).

8. *Newsweek*, 6 June 1977, pp. 87–88.

9. Paul Uselding, "Studies of Technology in Economic History," in *Recent Developments in the Study of Economic and Business History: Essays in Memory of Herman E. Krooss*, ed. Robert E. Gallman (Greenwich, Conn., 1977).

10. It is important, I think, to emphasize at the outset that even within armory practice—at least in the private sector—absolute interchangeability of parts was seldom attained. This, however, does not negate the characterization of armory practice as the pursuit of interchangeability through principles and practices of precision manufacture as I have outlined them. For additional information on interchangeability of parts, see my dissertation, cited in note 4.

11. Information on the early history of Wheeler & Wilson may be found in the following sources: Frederick G. Bourne, "American Sewing-Machines," in *One Hundred Years of American Commerce*, ed. Chauncey M. Depew, vol. 2 (New York, 1895), pp. 524–39; Grace Rogers Cooper, *The Sewing Machine: Its Invention and Development* (Washington, D.C., 1976), pp. 3–42; Wheeler & Wilson Sewing Machine Company, *The Sewing Machine: Its Origin, Introduction into General Use, Progress and Extent of Its Manufacture [and] A Great Machine-Shop Described* (New York, 1863); Carl W. Mitman, "Nathaniel Wheeler," *Dictionary of American Biography* 19:460–61; and "Wheeler and Wilson Manufacturing Company: Nathaniel Wheeler—A. B. Wilson," in J.D. Van Slyck, *New England Manufactures and Manufactories*, vol. 2 (Boston, 1879), pp. 672–82.

12. Joseph Wickham Roe pointed this out in his chapters on "The Colt Armory" and "The Colt Workmen—Pratt and Whitney," *English and American Tool Builders* (New Haven, Conn., 1916).

13. Guy Hubbard, "Development of Machine Tools in New England—Part Seventeen," *American Machinist* 60 (1924):877.

14. See Rosenberg's introduction to his *The American System of Manufactures* (note 2), p. 171.

15. Wheeler & Wilson Sewing Machine Company, *The Sewing Machine . . . [and] A Great Machine-Shop Described* (note 11), p. 16.

16. Production figures are taken from Bourne's article on the sewing-machine industry in *One Hundred Years of American Commerce* (note 11), p. 530.

17. Willcox & Gibbs and Brown & Sharpe histories may be found in the following sources: Cooper, *The Sewing Machine* (note 11); "Charles Henry Willcox," in *National Cyclopaedia of American Biography* 19:67–68; Henry Dexter Sharpe, *Joseph R. Brown, Mechanic, and the Beginnings of Brown and Sharpe* (New York, 1949); and Howard Francis Brown, "The Saga of Brown and Sharpe (1833–1968)" (Master's thesis, University of Rhode Island, 1971).

18. J.A. Willcox to J.R. Brown and Sharpe, 19 Mar. 1858. Historical Data Files

(Bound Volumes), Brown & Sharpe Manufacturing Company, North Kingstown, Rhode Island.

19. Henry Dexter Sharpe lists these tools in *Joseph R. Brown* (note 17).

20. J.R. Brown and Sharpe to James Willcox, 30 Oct. 1858. Historical Data Files, Brown & Sharpe. I am not aware of the identity of the Brown & Sharpe workman who had been employed at the Sharps rifle works.

21. For example see J.R. Brown and Sharpe to James Willcox, 7 Sept. 1858. Historical Data Files, Brown & Sharpe.

22. J.R. Brown and Sharpe to James Willcox, 21 Sept. 1858. Historical Data Files, Brown & Sharpe.

23. Production figures are from Bourne's article on the sewing-machine industry in *One Hundred Years of American Commerce* (note 11), p. 530.

24. Singer company history may be found in Cooper, *The Sewing Machine* (note 11); "American Sewing Machines" (note 11); John Scott, *Genius Rewarded; or the Story of the Sewing Machine* (New York, 1880); and Robert B. Davies, *Peacefully Working to Conquer the World: Singer Sewing Machines in Foreign Markets, 1854–1920* (New York, 1976).

25. Both the Boston *Almanac* (1854–56) and the Boston *Directory* (1853) list Phelps as a maker of philosophical instruments.

26. J.M. Emerson to I.M. Singer and Company, 24 Mar. 1851. Singer Manufacturing Company Manuscripts, State Historical Society of Wisconsin, Madison, Wisconsin, Box 189. For more information on early manufacturing at Singer, see my dissertation, cited in note 4.

27. See the contracts for this room and power in Singer MSS., Box 155. The hundreds of bills in Singer MSS., Boxes 231 and 234, reveal the sources and extent of Singer's materials, parts, and tool purchases.

28. *United States Magazine* 1 (15 Sept. 1854): pl. opp. 161.

29. Singer Manufacturing Company, *Mechanics of the Sewing Machine* (New York, 1914), p. 45.

30. I.M. Singer and Company [Edward Clark] to William F. Proctor, 16 July 1855, Singer MSS., Box 189. Proctor was a key employee of the company and later would be an officer in charge of production. Clark provided the business brains for the company from the formation of the partnership until his death in 1882.

31. For biographical information on Singer and Clark, see Carl M. Mitman, "Isaac Singer," *Dictionary of American Biography* 17: 188–89; "Isaac Singer," *National Cyclopaedia of American Biography* 30: 544–45; Peter Lyon, "Isaac Singer and His Wonderful Sewing Machine," *American Heritage* 9 (Oct. 1958): 34–39, 103–9; and "Edward Clark," in *Obituary Record of Donors and Alumni of Williams College, 1882–3*, No. 18 (1883): 304–6.

32. I.M. Singer and Company [Edward Clark] to William F. Proctor, 16 July 1855, Singer MSS., Box 189.

33. See Rosenberg's introduction to his *The American System of Manufactures* (note 2), pp. 2–18, and Robert F. Dalzell, Jr., *American Participation in the Great Exhibition of 1851*, Amherst College Honors Thesis No. 1 (Amherst, Mass., 1960).

34. B.F. Spalding, "The 'American System' of Manufacture," *American Machinist* 13 (1890): 11. For information on Singer's New York factory, see "Sewing Machines," in *Eighty Years' Progress of the United States . . . , by eminent literary men* (Hartford, Conn., 1866), pp. 421–23.

35. See the following letters: George B. Woodruff to I.M Singer and Company, 3 Mar., 14 Mar. (twice), 13 May, 20 May 1862, Singer MSS., Box 2; Jerome Carter to I.M. Singer and Company, 29 Sept. 1862, Singer MSS., Box 5; and power of attorney form, Mrs. Lydia M. Carter, administratrix of Jerome Carter to Singer Manufacturing Company, 19 Sept. 1876, Singer MSS., Box 190.

36. ASME *Transactions* 44 (1922): 927–28. The hiring of Miller may have been the occasion recalled by Isaac Singer, who still worked around the factory in 1863, when "his foreman nearly ruined the business by his first attempt to produce parts on the interchangeable plan without the exactness of duplication that was later obtained." "Looking Back: A Bird's Eye View of Our Trade," *Sewing Machine Times*, n.s. 18 (25 Oct. 1908):8.

37. This is an impression that I get from Waldo E. Nutter, *Manhattan Firearms* (Harrisburg, Pa., 1958).

38. I have reason to believe that Singer acquired this know-how from the Providence Tool Company, which from 1871 to 1873 manufactured for Singer a lower-grade sewing machine marketed under the name Domestic. See my dissertation, cited in note 4, pp. 132–40.

39. Robert B. Davies, "International Operations of the Singer Manufacturing Company, 1854–1895" (Ph.D. diss., University of Wisconsin, 1966), p. 124, citing Singer Directors' Minutes, 18 Feb., 19 Oct., 31 Oct. 1870, and 25 Jan. 1871. McKenzie eventually made all of the important decisions regarding manufacturing operations at Singer.

40. See production figures in Bourne, "American Sewing Machines" (note 11), p. 530, and in my dissertation (note 4), p. 117a.

41. "Some Recollections of the Brown and Sharpe Shop, South Main Street," Historical Data File (Bound Volumes), Brown and Sharpe Manufacturing Company.

42. Henry M. Leland to John R. Freeman, 11 Feb. 1927, Historical Data File (Bound Volumes), Brown and Sharpe Manufacturing Company. See also Henry M. Leland to Brown and Sharpe Manufacturing Company, 15 Apr. 1914; "Henry M. Leland conference with L.D. Burlingame, August 24, 1929" in Brown and Sharpe's Historical Data File (Bound Volumes); and Ottilie M. Leland, *Master of Precision: Henry M. Leland* (Detroit, 1966).

43. *Scientific American*, n.s. 41 (1 Nov. 1879): 271.

44. *Scientific American*, n.s. 40 (3 May 1879): 271.

45. John Scott wrote the book, and it was published in New York by John J. Caulon.

46. L.B. Miller to Singer Manufacturing Company, 3 Aug. 1885, responding to a questionnaire by the United States Bureau of Labor. Singer MSS., Box 200.

47. Scott, *Genius Rewarded* (note 45), pp. 48, 50.

48. Ibid., p. 50.

49. L.B. Miller to Singer Manufacturing Company, 3 Aug. 1885. Singer MSS., Box 200.

50. See Singer Manufacturing Company, *Catalogue of Singer Sewing-Machine* (1896) and Singer Sewing Machine Company, *Mechanics of the Sewing Machine* (1914).

51. This judgment may seem absurd, but if one studies carefully the development of Frederick Winslow Taylor's brand of scientific management it becomes clear that his approach to management grew out of a machine-shop/heavy-manufacturing environment rather than out of light-metal, consumer-durables production (which has been identified as the American system of manufactures). Taylor's own work and the early work of his disciples took place in manufacturing establishments operated on a different production system than that used, say, at the Colt factory in Hartford. A few American-system/armory-practice firms—notably Remington Typewriter and Winchester Repeating Arms—eventually hired Taylor consultants, but after 1910, very late in the development of the American system of manufactures and moderately late in the development of Taylorism. In his essay in this volume, Daniel Nelson suggests that American system workers had fewer problems than textile, machine-shop, or heavy-manufacturing workers. This certainly squares with my interpretation of Taylorism and the American system of manufactures. Taylor's work is best delineated in Frank B. Copley, *Frederick W. Taylor: Father of Scientific Management*, 2 vols. (New York, 1923). See also Daniel Nelson, "Scientific Management, Systematic Management, and Labor, 1880–1915," *Business History Review* 48 (1974): 479–500.

52. "Mass Production," *Encyclopaedia Britannica*, 13th ed., new vols. (1926 suppl.), 2: 821.

53. This episode is covered in detail in my dissertation cited in note 4, pp. 151–79.

54. A.D. Pentz to G.R. McKenzie, 2 Apr. 1881. Singer MSS., Box 193.

55. Ibid.

56. See, in this volume, Chandler, "The American System and Modern Management," and Nelson, "The American System and the American Worker."

57. John Buttrick, "The Inside Contract System," *Journal of Economic History* 12 (1952): 205–21. Felicia Deyrup discusses the inside contracting system in *Arms Makers of the Connecticut Valley*, Smith College Studies in History 33 (Northampton, Mass. 1948), pp. 160–64. See also Daniel Nelson, *Managers and Workers* (Madison, Wis., 1975), pp. 36–40, and Alfred D. Chandler, Jr., *The Visible Hand* (Cambridge, Mass., 1977), pp. 271–72.

58. Buttrick, "The Inside Contract System" (note 57), pp. 206–7.

59. A.D. Pentz to G.R. McKenzie, 13 June 1881. Singer MSS., Box 193.

60. A.D. Pentz to G.R. McKenzie, 29 June 1881. Singer MSS., Box 193.

61. A.D. Pentz to G.R. McKenzie, 19 May 1881. Singer MSS., Box 193.

62. Sydney A. Bennett to G.R. McKenzie, 1 June 1881. Singer MSS., Box 193.

63. Sydney A. Bennett to G.R. McKenzie, 19 July 1881. Singer MSS., Box 193. John Buttrick noted similar problems occurring with Winchester, "The Inside Contract System" (note 57), pp. 210–11.

64. For more information on Bennett, see his obituary in ASME *Transactions* 19 (1897–98): 979–80.

65. Minutes of a meeting held at Elizabethport factory, 26 Mar. 1883. Singer MSS., Box 239.

66. For more information on the work of this committee, see my dissertation (note 4), pp. 167–70.

67. A.D. Pentz to G.R. McKenzie, 12 Dec. 1884. Singer MSS., Box 198.

68. A.D. Pentz to G.R. McKenzie, 17 Aug. 1884. Singer MSS., Box 198.

69. Unfortunately, this bluebook does not survive. Details about it are to be found in correspondence between Bennett and McKenzie and also in "Minutes of meeting held at Kilbowie, June 26, 1886." Singer MSS., Box 204.

70. The correspondence between McKenzie, Bennett, and L.B. Miller, 1884–86, suggests that these officers of the company recognized problems inherent in the inside contracting system. But, as noted, piece-rate disputes were the immediate cause of the installation of the foreman system.

71. See my dissertation (note 4), passim.

THE AMERICAN SYSTEM AND MODERN MANAGEMENT

Alfred D. Chandler, Jr.

The American system of manufactures can be defined as a process of high-volume production by means of the fabrication of standardized parts that assemble into finished products. If this definition is acceptable, then there was a close relationship between the widespread adoption of the American system of manufactures and the emergence of modern industrial management. The system played this essential role because of both the complex nature of the process and the complex nature of the product. The complexity of process led to the creation of practices and procedures of modern factory management. That of the product helped to develop those of modern corporate management.

The American System of Manufactures and Factory Management

In order to evaluate the impact of the complexities of the process on the evolution of factory management, one must understand what differentiated the American system of manufactures from other new processes of high-volume production. Let us start then with a review of the beginnings of modern mass production in the United States.[1]

There were two preconditions. First, the railroad and telegraphic network had to be built in order to permit large amounts of raw materials to move at a rapid but steady pace into the factories, and the finished goods to be shipped out as regularly and as speedily. Second, there had to be available large quantities of fossil fuel (in this case coal) to provide the essential heat and power. The new production processes, therefore, had their beginnings in the 1850s as the railroad and telegraph networks expanded and as coal supplies became abundant. They reached fruition in the 1880s after the railroads had perfected cooperative arrangements to haul freight cars to any part of the country without transshipment and after the completion of the telegraph and cable network enabled factories to have almost instantaneous communication with any town in the country and indeed in many other parts of the world.

The new transportation and communication brought innovative large-batch and continuous-process techniques first to the refining and distilling industries. There the regular and rapid movement in processing of liquids through the stages of production was relatively easy to plan and design. Comparable high-volume methods soon appeared in industries based on mechanical processes, particularly those processing agricultural products. However, because of the complex nature of the process, mass production came more slowly in iron, steel, copper, and other metalmaking industries and those industries which transformed metal into implements, machinery, and other finished products.

In all these industries, rapid increases in the volume and speed of output and in the productivity of the worker, and decreases in the unit cost of production, resulted from three factors: improved machinery, better design of factory works, and more intensive use of energy. Only in the case of the metalworking industries—industries in which mass production was achieved by fabricating and assembling standardized parts— did expanding output and productivity and decreasing costs require close attention to a fourth factor: designing and managing the movements of men.

Innovation in the processing of oil provides a revealing example of the coming of mass production in the refining and distilling industries. By 1869, ten years after the discovery of oil at Titusville, Pennsylvania, refineries had been designed so that the flow of petroleum through the works, as it was converted from crude oil into kerosene and other products, was accomplished without human labor. Insofar as workers were needed, it was primarily to package the final product. In these early years, output was constantly expanded by improving stills and the steam-powered machines used to move the petroleum through the works. Just as important were the more intensive use of energy by the adoption of superheated steam distillation (borrowed from recent innovations in the refining of sugar) and the new "cracking" process, a technique of applying higher temperatures to reshape the molecular structure of crude oil. Comparable increases in production were achieved during the 1860s and 1870s in the processing of sugar, cottonseed oil, alcohol, and acids; in the distilling of whiskey and other spirits; and in the brewing of beer.

In the processing of agricultural products and in allied industries, a number of machines were invented in the late 1870s and 1880s to permit large-batch and continuous-process production. New machines made possible the mass production of cigarettes, matches, flour, breakfast cereals, canned foods and milk, and soap. The cigarette-making machine patented by James Bonsack in 1881 provides a representative example of such innovations. That machine, which produced 120,000 cigarettes a day as compared to an output of 3,000 by the most skilled workers, swept the tobacco into an "endless tape," compressed it into a rounded form, wrapped it with tape and paper, and carried it to a "converging tube," which shaped the cigarette, pasted the paper, and then cut the resulting rod into individual cigarettes. In the metalmaking industries the Bessemer and the Open Hearth techniques, both large-batch processes, made possible the mass production of steel. Both innovations relied on vast amounts of coal to provide the necessary heat. Their voracious appetites for pig iron as well as coal encouraged the integration of several processes of steelmaking within a single works—the blast furnace to make the pig, the Bessemer or Open Hearth to convert it to steel, and the rolling mill to produce rails, beams, and other finished products. Works design thus became a critical factor in the speed with which materials flowed through the stages of production. Alexander Lyman Holley, the brilliant engineer who built nearly all the early Bessemer works in the United States, designed them so that they greatly outproduced the leading British establishments. Once completed, output of the new works was increased by improving equipment and intensifying the use of energy. As Peter Temin has pointed out: "Steam and later electrical power replaced the lifting and carrying action of human muscle, mills were modified to handle the steel quickly and with the minimum of strain to the machinery, and people disappeared from the mills. By the turn of the century, there were not a dozen men on the floor of a mill rolling 3,000 tons a day, or as much as the Pittsburgh rolling mill of 1850 rolled in that year."[2]

In the same years that Holley was building the new Bessemer works, the American system of manufactures was increasingly used to mass-produce implements and machinery. In these mass-producing factories men did not disappear from the floor of the shops. Only in the second half of the twentieth century, with the coming of automatic machinery and "automation," did such metalworking factories begin to take on the organizational characteristics of nineteenth-century refining and distilling, food processing, and metalmaking industries. The metalworking factories did benefit greatly from constant improvement in machinery and plant design, and a more intensive use of energy. Indeed, the pioneers of modern factory management, such as Frederick W. Taylor, were mechanical engineers and designers who made advances in all these areas.

In metalworking factories, however, a large force using a wide variety of skills carried out many different tasks. The sewing-machine factory provides a good example. Charles H. Fitch's description in the introduction to the *Census of Manufacturers* for 1880 suggests the complexity of this type of production and notes the many different materials used, including "pig-, bar-, and sheet-iron, iron and steel wire, bar- and sheet-steel, malleable iron, japan varnish, and power and machine supplies in general, woods for casing (largely walnut and poplar), besides a considerable range of other materials."[3] In the making of metal parts, the bulk of materials passed successively from one operating department to another—from the foundry to the "tumbling-room, annealing, japanning, drilling, turning, milling, grinding and polishing, ornamenting, varnishing, adjusting, and proving departments." In addition, there were other metalworking departments producing tools, attachments, and needles. The "wood-working and cabinet-making departments constitute[d] a separate and distinct manufacture" probably as complicated as any massproducing furniture factory of the period. Finally, a large assembly department was responsible for the completion of the product and its gauging, inspection, and preparation for shipment. The coordination and control of the flow of materials at a high volume through many departments in which many workers were employed in each of the production processes created most challenging administrative problems in the nineteenth-century American industry.

These problems became acute, however, only in the later 1870s and 1880s. Before 1840, it must be remembered, the only use of interchangeable parts was in the production of small arms for the United States Army. Only the army could subsidize the high cost of moving materials into the Springfield and other armories and then moving the finished firearms out. And only it could pay the high cost of the necessary fuel. Then in the late 1840s and 1850s, as cheap coal became available and transportation and commuication improved, volume production by the fabricating and assembling of standardized parts became common in the making of locks, watches, clocks, sewing machines, and agricultural machinery, as well as the new breach-loading revolvers, rifles, and shotguns for commercial rather than military markets. Consumer demand for all these items continued to rise quite steadily until the onset of the depression that began in 1873. To meet this demand, the manufacturers, particularly in the years after the Civil War, concentrated on improving the technology of production.

In these years of expansion, manufacturers relied on skilled foremen to recruit, train, and manage the working force. Many of these foremen were themselves recruited from the Springfield Armory or from the early factories that made simple agricultural implements such as axes, hoes, and shovels. Increasingly the manufacturers relied on "inside contracting." By this system of labor organization, owners contracted with departmental foremen to produce a specified number of parts or mechanisms in a specified time (usually a year) for a specified price. The owners agreed to provide the floor space, machinery, light, heat, power, special tools, and patterns, and the necessary raw and semifinished materials. In addition, the contractor normally received a foreman's wage, which assured him of a minimum income. The contractor was, in turn, responsible for hiring, training, managing, and paying his labor force.

Under this system the owners were able to pass on to the contractor complicated problems of labor management. At the same time, however, they lost control over costs and the coordination of the flow of goods through the many departments. The contractor had every incentive to conceal information on costs from the factory owners, as such information affected his own bargaining position. Nor did the contractor feel the need to coordinate flows so as to use expensive machinery more continuously and efficiently. The company, not he, paid these machinery costs.

With the prolonged depression of the 1870s and the resulting drop in orders, manufacturers had to turn their attention to reducing per-unit costs and improving the coordination of flows through the plant. Henry C. Metcalfe, Henry A. Towne, Frederick W. Taylor, and others began to develop what Metcalfe, an army officer who had supervised several armories, called the shop-order system of accounts. Metcalfe, who began describing this system in 1885 in *The Cost of Manufactures and the Administration of Workshops, Public and Private* (the first book ever

Frederick Winslow Taylor (1856–1915). (Smithsonian Institution Neg. No. 61600–C.)

Henry A. Towne (1844–1924). (Smithsonian Institution Neg. No. 81–192.)

written in the United States on factory management), essentially adapted to the needs of interchangeable-parts manufacturing the voucher system of accounts developed in American railroad repair shops. When this method was perfected, each order accepted by the factory received a number which was then put on routing slips prepared at the plant office. On these slips each department foreman noted the time and wages expended, as well as the machines and materials used on an order while it was in his department. The ticket acted as an authority to do the work and to requisition material. It also became a "roll call" for the working force. From this basic system, sophisticated new techniques in inventory and quality control were developed. The execution of work under the system did, however, present two basic difficulties. First, it required a great deal of paperwork from the foremen; second, it called on the inside contractors for information they had little desire to provide.

Partly as a way to get the contractors and other foremen to accept the new "shop-order" system, Henry A. Towne of Yale and Towne, Frederick A. Halsey of the Rand Drill Co., and other metalworking manufacturers developed what they termed "gain-sharing" plans. These plans were, in fact, contracts with the foreman and his working force that retained incentives for efficient production and yet permitted the owners to obtain control over cost and coordination. Any reduction in unit costs achieved by improved machinery and plant design, better scheduling, or more productive labor would be shared equally by the company and the labor force. Of labor's share, sixty to eighty percent was to go to the foreman. Such plans proved acceptable. Gain-sharing replaced inside contracting.

In 1895, in a paper delivered at a meeting of the American Society of Mechanical Engineers, Frederick W. Taylor made his first pronouncement about "scientific management." He addressed himself directly to the gain-sharing plans of Towne and Halsey. These he criticized on two counts: first, the costs and the resulting savings to be shared were based solely on historical experience; second, the plans failed to include a stick as well as a carrot. Taylor insisted that standards on which bonuses and premiums were paid should be "scientifically," not historically, determined by such techniques as time and motion studies. For the stick, he proposed a "differential piece rate." Workers who failed to meet standard output received a lower rate per piece, just as those who excelled received a higher one.

Taylor's efforts to determine such standards left him convinced that the work of a whole shop or department was too complex to be carried out by a single foreman, no matter how skilled and well trained. He urged that the general foreman be replaced by a number of "functional foremen" and a "planning room." The workers on the floor of the shop or department

Stopwatch graduated in hundredths (ca. 1920), designed by Frank B. Gilbreth. (From the collections of the Division of Mechanical and Civil Engineering, National Museum of American History; Smithsonian Institution Neg. No. 81–144.)

would report to eight different foremen, including a route clerk, an instruction-card clerk, a cost-and-time clerk, a gang boss, a speed boss, a repair boss, an inspector, and a "shop disciplinarian." All these functional foremen would report to the planning room at the factory's central office. The planners would schedule flows, set the daily work plan for each department, and keep constant watch over costs. They would also be responsible for hiring, firing, and promotion of personnel, and for the maintenance or modification of existing standards.

American manufacturers were less than enthusiastic about Taylor's goal of extreme specialization in management. Very rarely did they adopt his full system. Its great defect was that it failed to pinpoint authority and responsibility for achieving departmental tasks and for maintaining a steady flow of materials from one stage of the process to another. To allocate responsibility for work and coordination and at the same time to benefit from the functional specialization that Taylor advocated, they instituted an embryonic line-and-staff organization. Individual departments or shops continued to be operated by foremen who were generalists and were on the line of authority and responsibility that ran from the president through the works manager and his direct subordinates. The specialized functions of Taylor's planning department and his functional foremen were assumed by the staff, who assisted and advised managers and foremen but had no direct control over operations. These changes came in an ad hoc manner, as managers and consulting engineers attempted to fit Taylor's ideas to the realities of managing day-to-day production. Henry Gantt, one of Taylor's most committed disciples, set up a new organizational structure at the Remington Typewriter Factory at Illion, New York, in 1910. By this plan of organization, all units involved in the fabricating and assembling of parts were managed through the manufacturing department—the de facto line department. Each shop within that department had a foreman responsible for its work and output. The heads of other departments—purchasing, stock order, shipping, inspection, time and cost, works engineering, and labor—became staff executives reporting directly to the works manager or his assistants and communicating with the factory departments through these senior line managers.

Two months before Gantt finished the reorganization at Remington, Harrington Emerson had completed a series of articles in *Engineering* magazine which defined more explicitly the role of line and staff in American factories. Emerson, a railroad man who had been troubleshooter for the Burlington and then the Santa Fe, fully understood the line-and-staff concept of business organizations—a concept which had been invented by American railroad men in the 1850s. In these articles, Emerson proposed four major staff offices for American factories—personnel, plant and machinery, materials, and methods and procedures. As with the railroads, staff executives were to set standards and supervise the small number of workers in their own specialized departments, but they were not to have responsibility for ordinary day-to-day operations. "It is the business of the staff," Emerson insisted, "not to accomplish work but to set up standards and ideals, so that the line can work more efficiently."[4]

With the acceptance of the line-and-staff type of organization, American factory management reached its modern form. In that same first decade of the twentieth century, A. Hamilton Church, Hugo Diemer, and other management experts had devised the sophisticated cost accounting so essential to determining profit and loss in complex manufacturing processes. Innovations came primarily in ascertaining indirect costs, or what was termed "the factory burden." Of particular importance were methods developed to relate overhead costs or burden to the fluctuating flows of materials through the manufacturing establishment. For a given factory capacity, unit costs fell as volume rose and rose as volume fell. By determining the standard costs based on a standard volume of, say, eighty percent of capacity, Church, Gantt, and others defined the increased unit costs of running below standard volume as "unabsorbed burden" and decreased unit costs above that volume as "over-absorbed burden."

During the twentieth century, American factories have grown larger, more productive, and more automated, but their basic organizational structure and the basic techniques for coordinating and controlling their operations have changed little since 1910. These basic forms appeared and were perfected in those industries employing the American system of manufactures in which mass-production techniques required a large, differentiated working force carrying out a wide variety of tasks. In such industries the design of the movement of men was even more significant than the design of the works, machines, or flow of energy in increasing productivity and output, and so in decreasing the unit cost.

The American System of Manufactures and Modern Corporate Management

In the twentieth century, the large industrial corporation has become such a powerful economic institution that corporate management has assumed more importance than factory management. During the last two decades of the nineteenth century, corporations came to manage many factories, and also many sales and purchasing offices, distribution and transportation facilities, mines and other sources of raw materials, and research laboratories. During the twentieth century, methods of corporate management (that is, the ways developed to organize, coordinate, and control the activities of widespead industrial empires) changed more dramatically than those of factory management. Nevertheless, the basic forms of corporate management used in the production and distribution of a single line of products had their beginnings in the 1880s and were perfected by 1910.

The modern industrial corporation—the forerunner of today's *Fortune* 500—first appeared when entrepreneurs using the new mass-production techniques began to distribute the output of their factories on a national and often international scale. Pioneering measures were necessitated, not from the complexity of production processes but from the complexity of the distribution activities. In those industries in which existing marketers—wholesalers and retailers in consumer goods and manufacturing agents in producers goods—were able to dispose of the manufacturer's products with speed and efficiency, producers rarely attempted to build extensive marketing enterprises. Both producers and distributors remained relatively small business enterprises. This was the case for those manufacturers using the American system of manufactures to produce watches, clocks, locks, and guns, or those using standardized parts in the making of steam engines and other heavy machines. When, however, manufacturers found existing channels inadequate to distribute the quantity of goods that could be produced, they then created large national and often international marketing and distributing networks.

And the companies requiring the most sophisticated sales organizations were those which made the most complex products, nearly always by the American system of fabricating and assembling standardized parts.

Thus, when the operating procedures for the new transportation and communication systems were perfected in the 1870s and the early 1880s, the large integrated corporation appeared, quite suddenly and dramatically. It came in those industries producing high-volume, semiperishable packaged products and in those processing perishable items for more than local markets. It became even more dominant in a third set of industries in which manufacturers produced complex, new, relatively expensive types of machines that—if they were to be sold in volume—required initial demonstration, installation, after-sales service and repair, and consumer credit.

A brief examination of the history of the industrial corporation in the first two types of industries helps to highlight the reasons why those in the third—those using the American system of manufactures—led the way in the development of modern distribution and in the management of large integrated business enterprises.

The new mass producers who no longer sold their semiperishable products in bulk but in small identifiable packages carrying brand names continued to use the wholesale jobber to distribute the goods. The manufacturers, however, now became responsible for scheduling the flow of goods through the wholesaler to the hundreds and thousands of retailers in the national market. They became responsible, too, for advertising their clearly identifiable products. The approximately five billion cigarettes which James B. Duke's American Tobacco Company distributed annually, for example, required speedy, steady deliveries, for in those days before cellophane wrappings, tobacco quickly grew dry and bitter. To expand the market for cigarettes—still an exotic tobacco product—Duke advertised massively throughout this country and much of the world. The enterprises which, like American Tobacco, integrated mass production with mass distribution in the 1880s included the makers of matches (Diamond Match), flour (the Washburn and Pillsbury companies), breakfast cereals (Quaker Oats), canned foods (Heinz and Campbell's soup), canned milk (Borden), soap (Procter and Gamble), film (Eastman Kodak), kerosene (Standard Oil), margarine and vegetable oil (American Cottonseed Oil), and paint (National Lead and Sherwin-Williams).

Nearly all the producers of perishable goods for the national market went one step further, bypassing the wholesaler altogether. Here the pioneers were the meatpackers, and of these the foremost innovator was Gustavus F. Swift. Swift realized that the rapid, regular flow of products that had to be consumed within three weeks after processing required more than just the newly invented refrigerated railroad car. It demanded a network of refrigerated warehouses for distributing meat to retail butchers, and offices from which salesmen could travel, advertising be arranged, and the scheduling of flows be carefully monitored. Five other firms quickly followed Swift's example—Armour, Cudahy, Morris, Hammond, and Schwarzchild & Sulzberger. They dominated their industry for the next half century. Brewers who began to sell in the national market in the 1880s, including Pabst, Schlitz, Blatz, Anheuser-Busch, and Moelin, created distributing networks comparable to that of the packers, though on a smaller scale. So did the growers and distributors of bananas, who merged at the turn of the century to become United Fruit. Wholesalers had been unable to handle these products because they could not provide the necessary refrigerated transportation and storage facilities. Nor could they manage the detailed and precise scheduling required to distribute the results of processing five to six million cattle a year or as many millions of bunches of bananas.

The high-volume makers of complex machinery also bypassed the wholesaler. However, they moved into marketing more slowly than did the producers of perishable products. At first nearly all sold through large jobbers. In almost every case the manufacturer discovered that the jobber was unable to do the selling tasks efficiently. He could not effectively schedule the flows of thousands of machines produced monthly, indeed often weekly. Nor did he have the technical know-how to demonstrate or install the machinery and, of more importance, to provide for continuing service and repair after sale. Very few had the financial resources needed to provide consumer credit so essential if the machines were to be sold in volume. This was true for recently invented light machinery such as sewing machines (Singer), complex agricultural machinery such as harvesters and reapers (McCormick and Deering), office machines such as typewriters (Remington), cash registers (National Cash Register), adding machines (Burroughs), and mimeograph machines (A. B. Dick). It was also the case for the makers of standardized heavy machinery, including dynamos and motors (General Electric and Westinghouse), telephone equipment (Western Electric), elevators (Otis), pumps (Worthington), boilers (Babcock & Wilcox), printing presses (Mergenthaler), radiators (American Radiator), shoe machinery (United Shoe Machinery), flour-milling machinery (Allis-Chalmers), and conveyor belts (Link Belt).

A few of the first group, such as Singer, integrated forward all the way, setting up their own retail stores. Most of the mass producers of light machinery, however, came to rely on the franchise dealer who sold their lines on a salary plus commission basis. Such dealer organizations were soon backed up by an elaborate network of branch offices which scheduled the flows to the dealers, supervised repair and service facilities (often recruiting and training mechanics for the dealer's shops), and provided dealers with credit necessary to permit customers to pay in monthly installments. The history of these earlier metalworking enterprises emphasizes that the automobile makers, the nation's best-known mass producers, were the beneficiaries rather than the innovators of modern distribution techniques as well as modern production methods.

The makers of producer or industrial goods—electrical machinery, pumps, elevators, and the like—did not, of course, deal with retailers. They set up branch offices with sales forces who sold directly to commercial and industrial concerns and so replaced the manufacturers' agents, who had sold on commission and normally handled the accounts of several producers. The manufacturers' agents simply did not have the knowledge and capabilities essentail to sell, distribute, and service these technologically advanced products. Nor did they have the credit facilities necessary to finance the sales of costly equipment. The large manufacturers acquired these financial resources either from retained earnings or by making alliances with investment bankers. Soon they were hiring, as salesmen, college graduates trained in mechanical and electrical engineering. By the turn of the century they were beginning to build research laboratories whose personnel worked closely with the salesmen, the design engineers, and the production managers in developing new products and reshaping old ones to meet customer needs.

Most enterprises that grew large by integrating forward into distribution and backwards into purchasing and often into producing their own raw and semifinished materials soon came to have much the same basic organizational structures. The most elaborate were those of companies that produced standardized machinery. All had at least four large functional departments—sales, production, purchasing of essential materials, and finance—each headed by a vice-president. But the operating departments of the machinery companies included more lines of products, and the sales departments, usually divided regionally, provided more services than did those of large enterprises in other industries. In addition, smaller departments carried on more complex activities than their counterparts elsewhere, essentially staff functions such as research and development, traffic, personnel, and legal affairs. In the most technologically advanced companies, particularly those involved with the production of electrical equipment, engineering or design departments were formed. These functional major departments in turn were organized on the principle of the line and staff, with the staff executives handling research, personnel, traffic, and legal matters.

The top management of large mass-producing enterprises included the vice-presidents in charge of the functional line departments, the president, and often a full-time chairman of the board. Normally these salaried executives made up the executive committee of the board of directors. That committee's function was to coordinate departmental activities, to monitor departmental performance, and to allocate resources for future production and distribution.

One of the earliest carefully designed management structures was that of the General Electric Company, formed in 1892 as a merger of two of the three largest electrical manufacturers of the day. At General Electric, the vice-presidents in charge of their departments operated through committees that included senior line subordinates, as well as staff executives. At their monthly meetings these committees covered a wide variety of topics. At sales meetings, for example, the members considered pricing, competitors' activities, market conditions, customer needs and concerns (particularly credit), and the processing of major orders. The manufacturing committee reviewed the regular factory reports on cost, inventory, and output, and discussed standardization of products and machinery, selection of plants for processing new products, and procedures for determining factory costs. In the latter regard, the committee paid close attention to the pioneering work done by Taylor and others in factory management. In addition to monthly committee meetings, the manufacturing and sales department held annual and later semiannual meetings in New York and Schenectady of all departmental managers, from the field as well as the central office. These two- or three-day conferences, with carefully planned agendas, provided a way to maintain personal communication between the growing ranks of the firm's middle managers who were specializing in a single function. Such channels were, in turn, supplemented by a flow of circular letters and bulletins emanating from headquarters and even greater torrents of statistics, reports, and letters moving from the field to Schenectady.

At General Electric close attention had to be paid to scheduling flows and controlling inventories. This was a challenging task, for even in 1895 the company had some 10,000 customers, from whom it received 104,000 separate orders.[5] As yet little thought had been given to calibrating production to carefully estimated market demand. Such forecasts had to wait at General Electric and other mass-producing machinery companies until the massive inventory crisis created by the sudden drop of demand in the postwar depression of 1920 and 1921. On the other hand, much attention was devoted to matching product design with customer needs. Salesmen, manufacturing executives, equipment designers, and managers from the laboratories were in constant touch in order to coordinate technological innovation with market demand and at the same time to develop products that could be produced at the lowest possible unit cost. In these technologically advanced enterprises, ideas as well as material flows had to be carefully coordinated.

As other American industrial enterprises grew to the size of General Electric and as their products became technologically complex, they borrowed most of GE's management policies and procedures. The chemical companies in particular looked to General Electric as a model. The structure built there became and still remains today a standard method of organizing a modern integrated industrial enterprise producing a single line of products.

International Implications

The organizational innovations stimulated by the American system of manufactures in both factory and corporate management had almost as great, if less direct, an impact on the European economies as they did on the economy of the United States. By the turn of the century they were enabling Americans to undersell Europeans in their own countries. But the "American invasion" involved more than just the sale of American products abroad. It soon led to a large direct investment in plants and personnel. As they were creating their sales networks at home, the new mass-producing, mass-distributing firms set up branch offices abroad. Soon, to meet the demand generated by these branch offices, they erected overseas factories and organized purchasing units to buy for these works. Often they bought and operated sources of raw materials. By World War I, the American multinational corporation had become a significant institution in the world economy. In 1914, direct investment abroad by American corporations was equal to seven percent of the Gross National Product. In 1966, the amount of direct investment abroad was precisely the same: seven percent of Gross National Product.[6]

The initial American challenge in Europe was spearheaded by American machinery firms. By 1914, as Table 1 indicates, of the forty-one large American firms with two or more plants abroad or with one plant and a raw-material-producing facility, almost half were manufacturing machines. As early as the turn of the century, American companies dominated European markets in sewing and office machinery, harvesters and reapers, elevators, shoe machinery, printing presses, pumping machinery, telephone equipment, radiators, and railway airbrakes. Only in electrical machinery did they have strong competition, which came from Germany. The Germans excelled in the production of nonstandardized heavy industrial machinery. But the Germans, the British, and the French were

Table 1 Multinational Companies in 1914

(American companies with two or more plants abroad or
one plant and raw-material-producing facility)

Groups 20 and 21: Food and tobacco

American Chicle	Coca-Cola
American Cotton Oil	H. J. Heinz
American Tobacco	Quaker Oats
Armour	Swift
British American Tobacco	

Groups 28, 29, and 30: Chemicals and pharmaceuticals, oil, and rubber

Carborundum	Sterns & Co. (drug)
DuPont	United Drug (drug)
Parke Davis (drug)	U.S. Rubber
Sherwin-Williams	Virginia-Carolina Chemical
Standard Oil of N.J.	

Groups 35, 36, and 37: Machinery, electrical machinery, and transportation equipment

American Bicycle	Mergenthaler Linotype
American Gramophone	National Cash Register
American Radiator	Norton
Chicago Pneumatic Tool	Otis Elevator
Crown Cork & Seal	Singer
Ford	Torrington
General Electric	United Shoe Machinery
International Harvester	Western Electric
International Steam Pump	Westinghouse Air Brake
(Worthington)	Westinghouse Electric

Others

Alcoa (33)	Eastman Kodak (38)
Diamond Match (39)	Gillette (34)

Source: Mira Wilkins, *The Emergence of Multinational Enterprise* (Cambridge, Mass., 1970), pp. 212–13, 216. The Group Numbers are those used in the Standard Industrial Classification of the United States Census.

unable to compete on equal terms, even when given tariff protection, with American firms using the American system of manufactures in production and the new methods they had perfected in distribution.

The reasons for the success of the American machinery firms seem clear. The large domestic market gave an initial advantage by assuring them of lower unit costs than the much smaller domestic markets could promise British, German, or French manufacturers. In such capital-intensive, high-volume industries, increased volume lowered unit costs, at least until plant capacity was reached. In addition, the extensive sales networks the American firms fashioned to sell their products in volume could provide essential marketing services with an efficiency that the smaller European firms often lacked. The low unit-cost and the trained sales force thus created powerful barriers to entry. Potential competitors had to sell at a loss until they reached a comparable volume. In the meantime, they had to devise ways of providing better and more dependable services and easier credit terms than already were available.

The history of American machinery firms that had their beginnings and grew large by employing the American system of manufactures is thus more than just a story of organizational innovation and business success. It may help to account for differences in the industrial development of the most technologically advanced modern economies. This is because the large enterprise that dominated markets in many lands also became responsible for promoting continuing research and development in specific technologies. New products and processes come from the constant interaction of sales, design, and research personnel within the enterprise. Therefore, although British and French customers benefited from low prices and good services from American companies, their technicians contributed little to innovations in process and product. Innovations in machinery and electrical equipment came instead from laboratories in Pittsburgh, Schenectady, Chicago, and New York. The trained managers and scientists in these laboratories continued to have close contact with their colleagues at major universities and technical schools such as Harvard and the Massachusetts Institute of Technology. In the United States that fruitful linkage between science and technology, and the practical application of both to human needs, was first forged in the machinery and electrical-equipment industries.

In Germany the comparable linkage first came in the chemical and then quite quickly in the electrical-machinery industries. During the 1870s a small number of chemical firms concentrated in the Rhine Valley developed a high-volume technology for the production of dyes and other chemicals that dramatically lowered unit costs.[7] These innovations permitted the firms to capture quickly not only markets in Germany, but also in

Britain, France, and the United States. These firms began in the mid-1880s to build their own worldwide marketing force of technically trained salesmen who demonstrated to textile manufacturers and other customers the proper way to use their dyes efficiently and with high quality control. By 1900 German firms completely dominated the dye and other organic-chemical industries throughout the world. In the same years they enlarged their research laboratories and strengthened ties with scientists and technicians in the faculties of the universities in which they had been trained. So chemical research became the dominion of industrial and university laboratories in Germany.

Such institutional linkages may help to explain why Britain and France missed out on what has been called the Second Industrial Revolution—those fundamental industrial developments which occurred in the late nineteenth century based on machinery, electricity, and chemicals. Industrialists in these two economically advanced nations were unable to compete in price and service with American machinery makers and German chemical- and electrical-machinery manufacturers. In Britain and France technically trained salesmen, production engineers, laboratory technicians, and university scientists were not coordinated through the instrument of the large business enterprise. There, universities rarely created large progressive departments of mechanical, electrical, and chemical engineering. There, few university-trained scientists and technicians spent their careers in industry.

I am uncertain about the impact the new German chemical and electrical-machinery industries had on the emergence of modern business enterprise and its management; but I am certain that the new metal-working industries—particularly the machine-making industries—played a major role in the development of modern management in the United States. The success of the new methods of machinery making in turn permitted American machinery to contribute to industrializing abroad (both in advanced and less-developed economies) and to making the processes of mass production by the fabrication and assembling of interchangeable parts a basic symbol of American culture. These accomplishments rest not only on the American system of manufactures, but also on the new American system of marketing created to distribute the products of this manufacturing technique.

Notes

1. The information and interpretations developed in this essay (except for the last section) are taken largely from my *The Visible Hand: The Managerial Revolution in American Business* (Cambridge, Mass., 1977), especially chaps. 8, 9, and 12.

2. Peter Temin, *Iron and Steel in the Nineteenth Century: An Economic Inquiry* (Cambridge, Mass., 1964), pp. 164–65.

3. Charles H. Fitch, "Report on Manufactures of Interchangeable Mechanism," in *Report on the Manufactures of the United States at the Tenth Census* (Washington, D.C., 1883), p. 548.

4. Harrington Emerson, *Efficiency as a Basis for Operations and Wages* (New York, 1911), p. 112. The line-and-staff distinction was developed by the senior managers of the large East-West trunk lines. J. Edgar Thomson of the Pennsylvania Railroad made the most substantial contribution to the concept. The need for a definition of authority and responsibility among the officers in charge of the different functions on a division—that is, the managers in charge of the movement of trains and traffic, the maintenance of way and rolling stock, the repair of track, bridges, and other permanent facilities, and the handling of accounts—came when a road grew large enough to be divided into two or more operating divisions. When this occurred, a central office was created in which middle managers became responsible for coordinating and monitoring the different functional activities of the two or more divisions. At that time, the president delegated his authority to the officers in charge of trains and traffic—the general superintendent at the lower management level.

   Because these executives were explicitly on "the line of authority" from the board and the president, they became known as line officers. The other functional officers became staff executives. The line officers were responsible for the day-to-day—and indeed minute-to-minute—operation of the road. Only they could order men to carry out activities involved in the movements of trains and traffic. They decided when and where tracks and bridges were to be repaired, locomotives and rolling stock were to be used, and financial reports and accounts were to be submitted. The staff officers set the standards for their departments and worked with the general and division superintendents in recruiting and promoting personnel. Of course, they also controlled work within their own shops or offices. Details of the development of the concept are given in Alfred D. Chandler, Jr., "The Railroads: Pioneers in Modern Corporate Management," *Business History Review* 39 (1965): 16-40.

5. Harold C. Passer, "The Development of Large-Scale Organization, Electrical Manufacturing Around 1900," *Journal of Economic History* 12 (1952):386.

6. Mira Wilkins, *The Emergence of Multinational Enterprise* (Cambridge, Mass., 1970), p. 201.

7. This story is told most fully in John J. Beer, *The Emergence of the German Dye Industry* (Urbana, Ill., 1959).

THE AMERICAN SYSTEM AND
THE AMERICAN WORKER

Daniel Nelson

T he census of 1880, compiled at the end of the first half century of the
"American system of manufactures," contained two classic essays.
The first was Charles Fitch's study of interchangeable manufacture in the
United States, a work familiar to all students of American technological
development and still, a century later, an indispensable source on the
American system. The second was Carroll Wright's report on the "Factory System in the United States." Unlike Fitch's report, which was
descriptive and positive in tone, Wright's essay was interpretive, even
polemical, and defensive. After a short history of the rise of the factory
system in England and America, Wright embarked on his major task: a
defense of the factory as a social institution. For more than three-quarters
of his report he rebutted charges that the factory degraded the worker.
"Abuses," Wright acknowledged, "have existed, but not equal to those
which have existed in the imaginations of men who would have us believe
that virtue is something of the past." Wright was especially emphatic in
rejecting the charge that the factory encouraged "intellectual degener-
acy." "Employment of the muscles in several operations instead of one,"
he retorted, "has nothing in it to elevate the faculties."[1] For the real as
opposed to the imaginary deficiencies of the factory, Wright counseled
employer-initiated labor reform. He concluded with a description of model
housing and other welfare plans.[2]

  Though the content of the Fitch and Wright reports demands careful
consideration, it is their juxtaposition, rather than their substance, that
is most revealing. The American system and the lot of the factory worker
were subjects of considerable interest to nineteenth-century observers,
and they continue to attract the attention of twentieth-century scholars.
Yet they were, and are, typically perceived as different concerns. The
Census Bureau officials obviously thought so in 1880; and recent analysts
of the American system and students of American labor history have
operated with minimal overlap, to put it mildly. However, there are
excellent reasons for this seemingly artificial dichotomy. The advent of
the American system created not only a distinctive approach to the man-

ufacture of machinery, but a distinctive type of factory. From a social perspective it fostered a near-ideal work environment, one that encouraged technical creativity, economic advancement, and social mobility. As a consequence, workers fortunate enough to find employment in such factories were on the whole a prosperous and contented group. The advent of the factory in other areas of manufacture had quite different effects. In heat-using industries, where technology also had a decisive influence on the work environment and the workers' lot, whatever positive impact production methods might have had was offset by harsh and dangerous working conditions. In technical industries, on the other hand, managerial discretion rather than technology played the decisive role. The results, in either case, were labor unrest, periodic eruptions of strife and violence, public controversy, and the charges that Wright took such pains to examine. Nineteenth-century labor reformers and more recent labor historians naturally have focused on the problems of factories and factory workers. Those problems were seldom found in plants operating under the American system. Hence the contrast apparent in the 1880 census. Fitch could speak positively about the American system and Wright defensively about the factory system without contradicting each other.

In the half-century after 1830 the American system spread rapidly in the United States, embracing the manufacture of guns, watches and clocks, locks, sewing machines, agricultural implements, business machinery, and, after 1860, locomotives, steam engines, and other heavy machinery.[3] During the same period the factory system spread from the textile industry to most other forms of manufacture, and a substantial majority of industrial workers became factory employees. Though the men and women employed in American-system plants were a relatively small fraction of the total—probably less than five percent of the factory labor force two generations after John H. Hall achieved interchangeability at Harpers Ferry—they deserve more attention than they have received.[4] In some respects their role in the American system was as notable and worthy of study as that of the machines they operated. Nevertheless, their relatively small numbers, the tendency of scholars to focus on single plants or industries—the gun industry for example—and, I shall argue, the outlook that the American system fostered among manufacturers, have tended to obscure the distinctive features of the relationship between the American system and the worker. Those features are most apparent if the American-system factories are viewed in the larger context of nineteenth-century industrialism. Certainly this is true if the precision-machinery plants are contrasted with three leading factory industries: iron and steel, representing the heat-using group; and cotton textiles and shoe manufacture, from the mechanical group.

The unifying characteristic of the factories of the American system was the pervasiveness of technologically defined social relations. Specialized machinery and sophisticated machine processes were the basis of the American system; they were likewise the basis of the social system that developed within such plants. They decisively affected the character of the work and, less often, the physical setting in which the work was undertaken. They were also the dominant factor in personnel and labor-management relations and in the web of relationships that existed among workers. Above all, they distinguished the lot of most armory, sewing-machinery, or agricultural-implement employees from other workers. The manufacturer who introduced the American system installed more than a series of machines and a system of machine production; he also made fundamental decisions regarding the treatment of his labor force, his relations with his subordinates, and, ultimately, his conception of social progress.

Among the most notable features of the American system was the relative neatness and cleanliness of the factory buildings. In his report Fitch noted such plants were "handsomely laid out, built and equipped," and a variety of observers, from the British Committee on Machinery to the most prominent American technical journalists, concurred.[5] Decades before steel, glass, and concrete construction, before the safety movement, and before government regulation made the dark satanic mill an anachronism, precision-machinery manufacturers had established what even by twentieth-century standards were reasonably attractive and healthy working conditions. The finest machinery plants were comparable to the model textile factories of the midnineteenth century; and, while sweeping generalizations are dangerous, it is probable that those at the opposite end of the spectrum offered employees a safer and cleaner working environment than the most modern iron mills or even the majority of custom machine shops of the same period.[6] In some cases this condition was a product of managerial design; Colonel Colt's Hartford armory was his most effective advertisement. But, as Felicia Deyrup notes, armory workers generally, not just Colt's employees, were "a favored class" who had "little to complain of."[7] In fact, Colt obtained his advertising at bargain rates. The character of the work, apart from any philanthropic impulse or nascent conception of industrial psychology, dictated hospitable conditions. Technical and commercial success, as Fitch noted, demanded that "neatness and order" be given the "highest consideration."[8]

But there was another factor that offset the inclinations of even the most careless, callous, or irrational manufacturer. This was the distribu-

A shop at the Colt factory, with slotting machines (foreground), lathes (front, left), and jigging or profiling machines (rear). (From Charles H. Fitch, "Report on the Manufactures of Interchangeable Mechanism" [1883]; Smithsonian Institution Neg. No. 81–205.)

tion of the labor force within the factory. Compared to other metalworking industries, precision-machinery manufacture required relatively large numbers of workers in departments and jobs that were by nature reasonably safe and clean, and relatively small numbers of workers in departments and jobs that by nature were dirty and/or dangerous. Typically twenty-five to fifty percent of the production employees of such a firm were machine-shop workers; and machine-tool and assembly operations together accounted for fifty to sixty percent of the total. By contrast, forge and foundry employees were seldom more than twenty-five percent of the plant labor force and usually much less than that.[9] Many manufacturers—the watch- and gunmakers, for example—did not use castings and had no foundries, unquestionably the unhealthiest and least pleasant area of the metalworking plant. The combination of a clean and well-ordered shop and a pattern of work that placed most employees in the cleanest and safest departments distinguished the precision-machinery plants from other ironworking factories, not to mention the heat-using industries generally.

If we assume that concern over working conditions was one factor that drew metalworkers to trade unions, the contrast is even more apparent. Beginning in the 1850s, skilled ironworkers, like the skilled employees of most heat-using industries, formed labor organizations. Of all these groups, the molders were the most assertive and organizable. United at least in part by a common understanding of their collective plight, they formed the most powerful and at most times the largest of the metal-trades unions of the second half of the nineteenth century.[10] Through the 1860s the majority of union molders were stove molders; but in the 1870s the balance swung to machinery molders, including employees of precision-machinery firms.[11] Though the stove molders continued to dominate the organization—a fact that led to considerable tension and made the union publications a poor source on the American system—machinery molders were by no means quiescent. Molders in precision-machinery firms were as aggressive as molders in other metalworking establishments. But they were usually a lonely and isolated minority. It was not that other workers in such plants tried and failed to organize—a pattern common to the mechanical industries. They typically showed little or no inclination toward overt collective activity. The foundry in a precision-machinery factory was often as disagreeable as foundries elsewhere, but the rest of the plant was distinctively different from other factories.

Robert Ozanne's history of labor-management relations at the McCormick Company is, for the period before the 1880s, largely a history of the McCormick molders, who constituted approximately ten percent of the labor force.[12] Ozanne concludes that the McCormick foundrymen successfully maintained their organization, resisted managerial encroachment, and waged numerous strikes in the 1860s and 1870s, a period when other McCormick workers showed no similar inclination. A union partisan, Ozanne is at a loss to explain the militancy of the ten percent and the apparent docility of the other ninety percent who labored under similar conditions, save for the environment of the foundry. Not until 1885–86, a year of labor upheaval, when shop conditions clearly had little effect on

Molding floor, McCormick foundry, 1885. (Courtesy of the State Historical Society of Wisconsin.)

worker behavior, did McCormick employees in other departments follow the molders' lead.[13]

If the American-system factories provided most machinery workers with superior working conditions, they also paid higher wages than most manufacturing industries.[14] High wages were a consequence of another feature of the American system in the nineteenth century that contradicts the simplistic equation of division of labor and mechanization with debasement: its dependence on a skilled labor force. The shift from artisan to factory production in gun- or watchmaking, for example, did not lead to the decline of the skilled worker, only to the decline of the artisan. The American system created a new spectrum of skills, and, compared to other factory operations of the midnineteenth century, it was a spectrum heavily weighted toward the skilled end. In part it was the need for toolmakers and other maintenance workers, in part the complexity and intricacy of the manufacturing process, and in part the need, in Fitch's words, for "great watchfulness and adaptation to varying conditions" that reconciled a minute division of labor with high skills and high wages.[15] Though the exact combination varied by industry and decade, the result was similar: a far larger blue-collar elite than existed in most factories.

While distinguishing between skilled, semiskilled, and unskilled workers in nineteenth century factories is an uncertain task, it is clear there were substantial variations between the machinery plants. Estimates from the 1870s suggest that anywhere from one-fourth to two-thirds of employees fell in the skilled, high-wage category. The agricultural-machinery manufacturers probably had the smallest proportion of skilled workers, the watchmakers the largest. Sewing-machinery plants probably represented the middle ground. These differences seem to have been most closely related to the number of workers in assembly operations. The assembly of a watch was a delicate, demanding process that required a large proportion of the factory labor force, perhaps one-third of the total. The sewing machine, also intricate, required one-fifth of the labor force. On the other hand, the rifle and pistol were less complex; one-tenth of the production workers could assemble the output of the factory.[16] Unfortunately, there is very little information about the heavy-machinery industries, the frontier of the American system in the late nineteenth century.[17] A locomotive was as complicated as a watch and required nearly as large a proportion of the labor force to assemble it, but it was hardly a delicate mechanism. Did manufacturers attempt to subdivide and mechanize the erecting process? Did they consider assembly-line procedures? If not, why?

In any case the contrast between the precision-machinery factories as a group and other industries is far more striking. Deyrup emphasized this point when she wrote that the armsworkers were "among the most highly skilled and highly paid of American workers."[18] Whatever the variations among the precision-machinery plants, they appeared alike, even interchangeable, when compared to other factory industries. In contrast to the twenty-five to sixty percent of American-system workers who were skilled and highly paid, perhaps twenty to twenty-five percent of the employees of an iron mill in the third quarter of the nineteenth century were highly skilled.[19] Most of the remainder, two-thirds or more of the total, were skill-less manual laborers who performed the myriad backbreaking tasks characteristic of the industry before the advent of mechanical materials-handling devices. Only ten percent of the employees of Pepperell Mills, a leading cotton-textile manufacturer, were skilled, while nearly eighty percent were low-paid machine tenders, a ratio that seems to have been typical of the textile industry in the nineteenth century.[20] The introduction of the factory system in the shoe industry created a more complex pattern. While specialist hand- and machineworkers gradually replaced artisan shoemakers after 1860, this change was, by the standards of other industries, neither very profound or rapid. Specialization and mechanization did not eliminate the need for "dexterity and judgment on the part of the operator."[21] Yet for reasons that have never been adequately explained, the workers' anxiety was profound. In no other industry did the issues of division of labor and mechanization agitate so many for so long.[22]

These differences between industries were most apparent in communities where precision-machinery manufacturers operated alongside other factories. Fortunately Howard Gitelman's studies of Waltham, Massachusetts, an industrial town with two famous employers, the Boston Manufacturing Company and Waltham Watch, provide a detailed view of one such nineteenth-century community. At first it is the similarities among the Waltham factories, their operations, and workers that stand out. Both companies prospered between 1850 and 1870; Boston Manufacturing doubled its labor force to 600 and Waltham Watch, founded in 1853, had 500 workers by 1870. In 1880 Boston Manufacturing had 1,200 workers, Waltham Watch 1,350. Both firms followed similar recruitment practices and both were benevolent employers by the standards of the day.[23] Above all, work at both plants was highly specialized. Boston Manufacturing and Waltham Watch employees were divided by function and product line. At Waltham Watch "3,700 discrete operations were required in the production of a single watch" by 1890.[24]

Yet Gitelman concludes that "differences between the companies were of more consequence than their similarities"—differences, he adds, that were largely a result of "such factors as competitive position and technologies employed."[25] The most important of those contrasts were the skills of the watchmakers and the economic and social advantages that arose from that fundamental distinction. Waltham workers always earned more than Boston Manufacturing Company workers of the same sex. During the 1860s, the only period for which Gitelman presents comparable figures, the differential was substantial: the average daily rate of all male Waltham employees was higher than the average daily rate of overseers at Boston Manufacturing, fifty percent higher by 1865.[26] In the following years the division between the two groups of workers remained a constant in the Waltham community. Despite rapid expansion at Waltham Watch and a continuous effort to refine and specialize production, the gap did not close. As Gitelman concludes, "Company employees certainly suffered no loss in status, nor did the company lose its reputation as the best place to work in Waltham, because of the increasing specialization of work."[27]

Another apparent difference between the American-system factories and other plants was a greater opportunity for economic and social mobility. Among machine designers and skilled mechanics there was substantial movement between firms and industries employing the American system; in many cases such moves involved a transition from blue-collar to white-collar or even entrepreneurial status that is hard to imagine in other manufacturing industries.[28] For the majority of workers, however, opportunity appears to have been closely associated with persistence.[29] In most nineteenth-century factories, the man who got ahead was the man who remained on the job. Although the evidence is scattered and inconclusive, it suggests that turnover rates were lower and mobility rates were higher for workers in precision-machinery manufacturing than for those in other industries. Clearly this was true at Waltham Watch and Winchester Arms; it was probably true at Remington and Singer Sewing Machine; and very likely it was the situation at the agricultural-machinery firms.[30] Carroll Wright was probably correct when he asserted in his 1880 essay that the factory offered the worker a chance for a higher income, but he might have added that the American system offered the worker a good deal more. And it was an offer that frequently evoked a positive response.

A distinctive opportunity of the precision-machinery factory was the chance to become an inside contractor, a blue-collar entrepreneur. A response to the technological and managerial demands of precision work, the inside contract system was a concomitant of the American system throughout the nineteenth century.[31] The contractors were few in number, probably no more than 1,000 at any time; realistically, the shift to salaried foremen increased the mobility opportunities for production workers. But the possibility of becoming a contractor was only one advantage of the system. The contractor's presence had important intangible effects that helped distinguish the machinery factory from others. The contractor reconciled the old and new technologies, the artisan tradition and specialized high-speed machinery. His role indicated to fellow workers that factory labor was not synonymous with wage labor and that specialization was not inconsistent with technical creativity or a degree of autonomy usually found only in an independent enterprise. For the contractor's immediate employees, the system also provided tangible benefits: competent instruction and supervision, a personal relationship with the "employer," and immediate consideration of grievances and personal difficulties. Though craftsmen in other industries—iron, metals, glass, and pottery, for example—enjoyed immense powers by modern standards, they could seldom compare with the factory entrepeneurs of the precision-machinery firms.[32] Their authority was less because their tasks were less demanding.

Finally there is one other effect of the American system that warrants special consideration. If the benefits of the technology adopted in the machinery factories were as pervasive and, in retrospect, as obvious as I have suggested, why were contemporary observers oblivious to them?

The technical achievements of the American system won wide acclaim; why were the social achievements disregarded? In part the answer was the relatively small number of factories and workers and the apparent diversity of industries that were involved. But a more important factor was the incidental character of the positive relationship between the American system and the worker. Manufacturers introduced precision methods and interchangeability for technical and economic ends, not to manage workers or subdue discontent. And employees reflected their comparative good fortune by working productively and, in most instances, contentedly. It is hardly surprising, then, that manufacturers were slow to recognize the social impact of the American system. Most of them never commented on labor issues of any kind, much less the connection between technology and the generally superior social climate of their shops. By the 1870s, however, precision-machinery manufacturers, far more than other employers, subscribed to a series of fundamental assumptions about workers and their treatment.[33] Of these, two were of particular importance: the notion that workers were "economic men"; and the idea that informal, personal, and above all unstructured relationships were essential to harmony and cooperation.[34] To understand the importance of this development it is necessary to turn to industry generally and the rise of the "labor problem."

As suggested above, few industries enjoyed the inherent social advantages of the machinery plants. In consequence, most factory operatives toiled under more adverse circumstances or were subject to managerially defined social relations. Workers in the metal and other heat-using industries generally suffered the former handicap, while workers in the mechanical industries, a majority of the total, operated in a social environment that was in large measure a product of the deliberate actions of managers and supervisors. In both types of plants a rough division of labor existed between the plant manager and the first-line supervisor. The manager reserved for himself the determination of wage levels, the length of the work day, general conditions of labor, and the resolution of problems that interrupted routine. But he delegated day-to-day production and personnel matters to first-level supervisors. In retrospect, this arrangement virtually insured a high level of hostility, suspicion, and unrest. The manager established conditions that profoundly affected the employees despite little knowledge of their situation or concerns, and the supervisor compounded the problem with a potent mixture of favoritism and tyranny.[35] There were, of course, many variations on these themes. In industries in which the top manager's power was great for whatever reason, unrest usually took covert forms. In industries in which that power or aggressiveness was less apparent, open collective activity by workers was more often the result.[36]

As a result the factory perpetuated the contentious relationships that had characterized preindustrial manufacturing. The first half-century of the factory was a half-century of widespread unrest, a fact that the vicissitudes of the business cycle, union organizing, reform, and war tended to obscure.[37] The advent of "industrial discipline," while possibly making industrial workers more conscious of the timeclock, not to mention their new mass-produced Waltham watches, did little to make them more agreeable or subservient to their employers. Nor did the vaunted advantages of American society—high wages, the promise of mobility, or the absence of a rigid preindustrial class system—immunize them against strikes, violence, the blandishments of trade-union organizers, or a sense of class consciousness. In nearly all areas of manufacturing, not to mention transportation and mining, the advent of modern industry was associated with a more or less permanent state of antagonism and strife.[38]

Yet if the manager set the tone for labor relations and labor policies in the typical factory, he could also change it. And when it became apparent that the traditional approach to the worker had become ineffective, even counterproductive, he did change it. In the last third and especially the last quarter of the nineteenth century, small groups of manufacturers, mostly in the textile industry, launched a broad-gauge attack on the labor problem.[39] Or, more appropriately, they attacked some of the manifestations of the labor problem. Like Carroll Wright, their foremost intellectual ally and apologist, they believed that industrialism was a positive force in society and that only certain tangential details required alteration. Because of this approach—I would go so far as to say because of their unfamiliarity with the operation of the American system—their progress was slow, uneven, and in many instances transitory. It featured substantial initiatives in the realm of "fringe benefits" but little or no attention to the relationship between work and unrest, particularly to the role of the supervisor. Still, their activities left a permanent imprint on the evolution of employer-initiated labor reform and the personnel management movement.[40]

The machinery manufacturers, who might have decisively influenced these efforts, in fact rejected them. Their approach to their employees seemingly had been similar to that of other manufacturers. But the inadvertent effects of the American system of manufactures produced a different result and a different perception of the situation. Accordingly, they remained oblivious to the labor problem until the 1880s and then emphasized the social benefits of centralized systematic production management rather than, and even as an alternative to, labor reform.[41] As a consequence, their achievements and the lessons that might have been drawn from them played no part in the effort to redefine the relationship between managers and workers in the factory.

An extraordinary incident from a later period illuminates this point. At the turn of the century, the model factory of the National Cash Register Company was one of the best and most successful examples of the American system in operation. But John Patterson, the company president, had also become the nation's leading practitioner of employer-initiated labor reform. Under his guidance NCR provided every benefit common to contemporary welfare work, from medical facilities to a cooking school. Patterson clearly was an iconoclast, his factory an anomaly. In June 1901 several groups of NCR workers, led, typically, by the molders, struck the plant. Patterson responded in the customary fashion. He locked out his employees, broke the strike, and subsequently reevaluated and discarded many of his labor programs. It was a traumatic experience.[42] Yet, judging by the public statements of Patterson's peers, machinery manufacturers delighted in his plight. Frederick A. Halsey—assistant editor of the

The National Cash Register Company's "Machinery Hall," 1898. Note the neckties and white-collar shirts. (Courtesy of the National Cash Register Co.; Smithsonian Institution Neg. No. 81–194.)

American Machinist, creator of a popular incentive wage plan designed to eliminate labor unrest, spokesman for the simplistic view of labor problems common among machinery manufacturers, and a curmudgeon to boot—could barely suppress his glee:

> Belief in Mr. Patterson's entire sincerity and sympathy for him ... do not, however, change my belief that ... a great part of his work is, and was bound to be, null, void and of no effect. In this world men work for pay. The Saturday night pay envelope contains the tangible results of a week's toil. The natural desire of every man is first to increase the contents of that envelope and second to shorten his hours of work. What to him is the cooking school, etc. compared with an increase in the contents of his envelope?[43]

Halsey's statements were symptomatic of the state of mind that had developed among men associated with the American system over the preceding sixty years. It was a state of mind reflected in the strange juxtaposition of the Fitch and Wright essays in 1880 and in frequent later condemnations of "soup house philanthropy."[44] Above all, it was a state of mind which more than any other single factor explains the failure of nineteenth-century manufacturers and twentieth-century scholars to recognize or appreciate the inadvertently positive social impact of the American system of manufactures.

The deficiencies of the late-nineteenth-century reform movement are beyond our influence, but the experience of the precision-machinery manufacturers may yet help to redefine the ties connecting technology, management, and labor in historical writing. Though a new and more satisfactory synthesis must await more detailed examinations, two conclusions emphasized here may prove helpful. First, it is apparent that there was no inevitable relationship between the division of labor and mechanization and the congeries of difficulties that manufacturers called the labor problem. In the American-system factories, where these characteristics of nineteenth-century factory production were most apparent, unrest was minimal. In other plants—the shoe factories of eastern Massachusetts are perhaps the best example—there was far less specialization and mechanization but far more discontent. "Employment of the muscles in several operations instead of one," as Wright concluded, was not necessarily the key to happiness, on or off the job. This leads to the second conclusion, the crucial role of the managers in the nineteenth-century factory. In retrospect it is clear that the practice of management was conducive to friction and antagonism. Where technology interfered with that practice—in the American-system factories to a greater and in the heat-using industries to a lesser degree—the problem was limited or often contained through institutional mechanisms, for example, the internal contract system and the craftsmen's unions of the iron industry.[45] In other industries, in which production methods had less impact, the managers' realm was larger. It was primarily in those industries that the labor wars—to use Sidney Lens's term—exacted their toll.[46] Thus it is not surprising that machinery manufacturers became smug and complacent or that their complacency was reflected, albeit indirectly, in Fitch's account of the American system. Nor should it surprise us that enlightened textile manufacturers were anxious and defensive, a defensiveness that was reflected in Wright's essay. The tragedy was that they did not learn more from one another. But surely that failure is no reason to permit the dichotomy that characterized nineteenth-century industry to characterize twentieth-century scholarship as well. Our subjects and readers deserve a better fate.

Notes

1. Carroll D. Wright, "Report on the Factory System of the United States," U.S. Department of the Interior, Census Office, *Tenth Census, 1880, Manufactures* (Washington, D.C., 1883), pp. 17, 20, 34.

2. Ibid. James Leiby, *Carroll Wright and Labor Reform* (Cambridge, Mass., 1960).

3. David Hounshell has shown that this statement of the "conventional wisdom" is accurate only if confined to a handful of sophisticated and innovative firms. See David Allen Hounshell, "From the American System to Mass Production: The Development of Manufacturing Technology in the United States, 1850–1920" (Ph.D. diss., University of Delaware, 1978) and his essay in this volume.

4. See Wright, "Report on the Factory System" (note 1), pp. 9–14.

5. Charles H. Fitch, "Report on the Manufactures of Interchangeable Mechanism," in *Report on the Manufactures of the United States at the Tenth Census* (Washington, D.C., 1883), p. 61; Nathan Rosenberg, *The American System of Manufactures* (Edinburgh, 1969), p. 194; "Brown & Sharpe Manufacturing Company," *Scientific American* 41 (1 Nov. 1879): 273.

6. For examples of model textile factories, see Rosenberg, *The American System* (note 5), pp. 318-25. Also Fred H. Colvin, *Sixty Years With Men and Machines* (New York, 1947), pp. 20-29.

7. Felicia Deyrup, *Arms Makers of the Connecticut Valley* (Northampton, Mass., 1948), pp. 163-64. For Colt's plant see J. L. Bishop, *History of American Manufacturing* (New York, 1966), 2: 737-42; and Charles T. Haven and Frank A. Belden, *A History of the Colt Revolver* (New York, 1940), pp. 352-62. For the E. & T. Fairbanks Company, which adopted a similar policy,

see J. D. Van Slyck, *New England Manufacturers and Manufacturing* (Boston, 1879), 1:254.

8. Fitch, "Interchangeable Mechanism" (note 5), p. 61.

9. These figures are based on data in Fitch, "Interchangeable Mechanism" (note 5).

10. See Frank T. Stockton, *The International Molders Union of North America* (Baltimore, 1921), p. 23; also Jonathan Grossman, *William Sylvis, Pioneer of American Labor* (New York, 1945); and Daniel J. Walkowitz, "Statistics and the Writing of Working Class Culture: A Statistical Portrait of the Iron Workers of Troy, New York, 1860-1880," *Labor History* 15 (Summer 1974): 427.

11. Stockton, *Molders* (note 10), pp. 42-45, 48-50.

12. Robert Ozanne, *A Century of Labor Management Relations* (Madison, Wis., 1967), chap. 1; Ozanne, *Wages in Theory and Practice* (Madison, Wis., 1968), chap. 2.

13. Ozanne, *A Century* (note 12), chap. 1. David Hounshell concludes that McCormick did not achieve interchangeability until the 1880s and that the company employed large numbers of skilled employees for assembly and fitting operations. See Hounshell, "From the American System to Mass Production" (note 3), chap. 3. There was a Machinists and Blacksmiths union that flourished in the 1860s and 1870s, but apparently its members were mostly small machine-shop or railroad employees. See David Montgomery, *Beyond Equality: Labor and the Radical Republicans, 1862-72* (New York, 1967).

14. There is no systematic study of wages in machinery manufacture. However, see Clarence Long, *Wages and Earnings in the United States 1860–1890* (Princeton, 1960), especially pp. 69-77, and Stanley Lebergott, *Manpower in Economic Growth* (New York, 1962), pp. 543-47.

15. Fitch, "Interchangeable Mechanism" (note 5), p. 37. At Baldwin Locomotive the labor force was divided in the following way between 1865 and 1876: 38% skilled, 20% semiskilled, 42% unskilled. Pennsylvania Department of Internal Affairs, *Industrial Statistics* (1865-76), 4: 546.

16. Fitch, "Interchangeable Mechanism" (note 5), pp. 5, 35, 49, 62.

17. Ibid., pp. 48-59. W. Paul Strassman, *Risk and Technological Innovation* (Ithaca, N.Y., 1958), p. 130.

18. Deyrup, *Arms Makers* (note 7), p. 4.

19. See Jesse S. Robinson, *The Amalgamated Association of Iron, Steel and Tin Workers* (Baltimore, 1920), p. 19; Pennsylvania Department of Internal Affairs, *Industrial Statistics* (1865–76), 4:603.

20. "Pepperell Mnuf'g Co.," Pepperell Manufacturing Company Papers, E-H 1, Baker Library, Harvard Business School; Thomas Dublin, "Women, Work, and Protest in the Early Lowell Mills: 'The Oppressing Hand of Avarice Would Enslave Us'," *Labor History* 16 (Winter 1975): 114.

21. "The Boot and Shoe Industry as a Vocation for Women," U.S. Bureau of

Labor Statistics, *Bulletin* 180 (1915): 41. This is the best source on occupations in the shoe industry. For a broader survey see Blanche Evans Hazard, *The Organization of the Boot and Shoe Industry in Massachusetts Before 1875* (Cambridge, Mass., 1921). Also Alan Dawley, *Class and Community: The Industrial Revolution in Lynn* (Cambridge, Mass., 1977).

22. See John P. Hall, "The Knights of St. Crispin in Massachusetts, 1869–1878," *Journal of Economic History* 18 (1958): 161-75; Paul Faler, "Cultural Aspects of the Industrial Revolution: Lynn, Massachusetts Shoemakers and Industrial Mobility, 1826-60," *Labor History* 15 (Summer 1974): 367-94; John T. Cumbler, "Labor, Capital and Community: The Struggle for Power," *Labor History* 15 (Summer 1974): 395-415; Dawley, *Class and Community* (note 21).

23. Howard Gitelmen, *Workingmen of Waltham* (Baltimore, 1974), pp. 40, 64-74. Also see Gitelman, "Occupational Mobility Within the Firm," *Industrial and Labor Relations Review* 20 (Oct. 1966): 50-63; Gitelman, "The Labor Force at Waltham Watch During the Civil War," *Journal of Economic History* 25 (June 1965): 214-44.

24. Gitelman, "Labor Force" (note 23), p. 230.

25. Gitelman, *Workingmen of Waltham* (note 23), p. 11.

26. Gitelman, "Labor Force" (note 23), p. 224.

27. Ibid., p. 230.

28. Joseph W. Roe, *English and American Tool Builders* (New Haven, Conn., 1916), pp. 139, 187; Cyrus H. McCormick Diary, 6 May 1880, McCormick Company Papers, State Historical Society of Wisconsin, 4C, Box 1.

29. Gitelman, "Occupational Mobility" (note 23).

30. Ibid.; Harold F. Williamson, *Winchester* (Washington, D.C., 1952), pp. 90-91; "American Industries—No. 75: The Firearms Manufacture," *Scientific American*, n.s. 45 (3 Sept. 1881): 148; Henry Roland, "Six Examples of Successful Shop Management, V," *Engineering Magazine* 12 (Mar. 1897): 997-98; Ozanne, *Wages* (note 12), pp. 16-18, 30.

31. John Buttrick, "The Inside Contract System," *Journal of Economic History* XII (Summer 1952): 205-21; Roland, "Six Examples" (note 30), pp. 994-1000; Williamson, *Winchester* (note 30), chap. 7; Deyrup, *Arms Makers* (note 7), pp. 101-2; Alden Hatch, *Remington Arms in American History* (New York, 1956), pp. 68-69.

32. See Daniel Nelson, *Managers and Workers: Origins of the New Factory System in the United States, 1880–1920* (Madison, Wis., 1975), chap. 3.

33. See Monte Calvert, *The Mechanical Engineer in America, 1830–1910* (Baltimore, 1967), chap. 9; Edwin T. Layton, Jr., "American Ideologies of Science and Engineering," *Technology and Culture* 17 (Oct. 1976): 688-701.

34. See Nelson, *Managers and Workers* (note 32), chap. 3.

35. Ibid.

36. Cf. James Holt, "Trade Unionism in the British and American Steel Indus-

tries, 1880–1914: A Comparative Study," *Labor History* 18 (Winter 1977): 5-35, and Melton Alonza McLaurin, *Paternalism and Protest* (Westport, Conn., 1971), with the Faler, Crumbler, and Dawley works on the shoe industry.

37. See Norman Ware, *The Industrial Worker, 1840–60* (Boston, 1924) and Montgomery, *Beyond Equality* (note 13).

38. See H. M. Gitelman, "Perspectives on Industrial Violence," *Business History Review* 47 (Spring 1973): 1-23.

39. Nelson, *Managers and Workers* (note 32), chap. 6.

40. Ibid.

41. The outstanding example of this tendency was Frederick W. Taylor's system of "scientific management."

42. Daniel Nelson, "The New Factory System and the Unions: The NCR Dispute of 1901," *Labor History* 15 (Spring 1974): 163-78.

43. Frederick Halsey, "The National Cash Register Company's Experiment," *American Machinist* 24 (20 June 1901): 688-89.

44. Louis Bell, "The Philanthropy of Self-Help," *Cassier's Magazine* 22 (Sept. 1903): 444.

45. See Robinson, *Amalgamated Association* (note 19) and Holt, "Trade Unionism" (note 36).

46. See Sidney Lens, *The Labor Wars* (Garden City, N.Y., 1973).

❁❁❁❁❁❁❁❁❁❁❁❁❁❁

THE DRAMA OF CONSUMER DESIRE

Neil Harris

In a publication devoted to the American system of manufactures, this
essay may seem out of place. For it examines neither the subject nor
the time-frame which dominates most of the essays. Its focus, instead, is
turned away from the producer of goods, and toward the consumer.

Descriptions of nineteenth-century industrial growth have considered a
broad range of issues, including labor scarcity, patterns of imitation,
primacy of innovation, and economies of scale. Studies of production and
distribution abound. But the role and character of the retail buyer remain
mysterious. Works are now in progress, or have been recently completed,
which promise correction. As part of this enterprise I would like to examine a significant accompaniment to the expansion of the American system
of manufactures—the creation of an American system of consumption, the
establishment of relationships between Americans and a range of objects
that were unprecedented in number and variety. These relationships
were supported by a set of shopping rituals, and by habits of appropri-
ation, which have become the commonplaces of modern capitalist society.
They developed throughout the world. The pattern I present may or may
not be applicable to Britain, France, or Germany during the same period.
It is too early to engage in comparative speculation, and I wish to make
no special national claims. But the process of identifying, describing, and
evaluating the consumer, as a social type, proceeded steadily in the
United States from the late nineteenth century on.

I am concerned less with the objective history of the consumer, and
more with consciousness of the consumer's role. For that reason my
principal sources will be works of fiction which placed the buying process
within the social experience. My procedure has four steps. First of all, a
brief summary of the special national problems, primarily ideological in
character. Second, some reference to the initial age of consumer consciousness in fiction, coming at the end of the last century. Third, a
description of trends in the early twentieth century, in particular during
the 1920s, which intensified American object consciousness. And finally,

some examples of the way in which mass distribution affected the novelistic sensibility of the period. My argument, tout court, is that mass-produced objects and object relationships came increasingly to enter the American novel, both as symbols and as experiences. But the very standardization which produced such triumphs for the manufacturing and distributing systems is here defined as a major problem for the creative imagination. Thus a national style of purchasing began to be adumbrated by our literary figures, and fixed upon as a cultural metaphor.

If the American experience was not unique, it did offer, so far as historians of consumption are concerned, special circumstances. Public rhetoric, as early as the seventeenth century, revealed such awareness. Moralists—political leaders and religious advisers both—found that material prosperity posed grave problems for the American colonists. Traditions of suspicion toward the effects of comfort and security were especially powerful in New England and Pennsylvania. The fine edge of religious suspense was, after all, honed on a collective sense of insecurity. Puritanism, and radical Protestantism generally, rested upon an introspective obsession with determining the relationship between sign and essence, sanctification and justification; any form of relaxation could be dangerous. Wealth bred relaxation; it induced in sinners false confidence and lessened their sense of dependence upon divine government. Economic success might spell spiritual failure, and the Puritan ministry relentlessly explored this paradox.[1]

It was also a problem for political leaders of the late eighteenth century, sensitive to charges of mercenary ambition. Republicanism seemed to demand austerity and restraint in personal life; the new task was to inspire the economic energies that aided national advancement and independence, without encouraging self-enhancement to assume a motive power all its own. That delicate jewel, virtue, without which the constitutional mechanism was doomed, could be marred by excessive appeals to personal advancement, yet qualities of discipline, industry, ingenuity, and audaciousness had economic gain and a higher standard of consumption as their almost inevitable epilogue.[2]

The problem did not end in the American Enlightenment. If anything it assumed grander proportions during the Jacksonian era. Jeremiads multiplied warnings of the dangers inherent in personal wealth and the traps to virtue and piety that lay concealed beneath fancy clothes, large houses, silver plate, and fine furniture. One American historian, analyzing the literature, perceptively labelled the psychic drama of the Jacksonian period "Prosperity the Riddle."[3] There were, of course, always watchmen on the walls quick to spot any outbreaks of the materialistic virus which could bring on fatal attacks of self-indulgence. Busybodies, who have excited the scorn of native satirists and foreign critics alike, with their burrowing curiosity about the personal tastes and lifestyles of private citizens, were nurtured in this hothouse of anxiety. What people drank, ate, wore, and consumed, seemed to have direct relationships to their laws, their rulers, and their morals.[4]

But while there are continuities between pre- and postmodern fears of personal consumption, there are also contrasts. In general the buying habits which attracted criticism through the middle of the nineteenth century involved the purchase of luxuries and the search for sensual pleasure. The upholsterer, the parfumier, the milliner, the jeweler, all had benefitted for centuries from the patronage of the wealthy and the aspiring. And the national concern that a new aristocracy, eager to demonstrate its aesthetic tastes and well-bred gentility, would be a public danger by reason of its frivolity and social competition, this concern had already been voiced.

Yet there were also some special fears voiced. It was not only that the American experiment involved self-restraint; Americans seemed particularly vulnerable to a new kind of self-indulgence. One need turn only to that vade mecum of cultural insights, Alexis de Tocqueville's *Democracy in America*, to get a sense of the urgency. Tocqueville discovered, defined, and discoursed on American materialism. He found its blessings mixed. A concern with physical improvement encouraged a series of constructive pieties and useful disciplines, Tocqueville acknowledged, but in this country it also constrained the role of idealism and the possibility for heroic acts. And the reason lay in the principle of equality. "The reproach I address to the principle of equality," Tocqueville wrote, "is not that it leads men away in the pursuit of forbidden enjoyments, but that it absorbs them wholly in quest of those which are allowed." The problem of American materialism—or its opportunity for American manufacturers and retailers—was not that it would corrupt a small number of community leaders, but that it would enervate, distract, and dominate the great masses. "The enjoyment of others is sanctioned by religion and morality; to these the heart, the imagination, and life itself are unreservedly given up, till, in snatching at these lesser gifts, men lose sight of those more precious possessions which constitute the glory and greatness of mankind."[5]

Tocqueville's generation was concerned with what has since come to be defined as the rise of mass society, and the conservative predilections of so many early sociologists helped shape much of their research and their writings. Later generations of critics would pick up and expand their message. American materialism was dangerous precisely because it did not confine itself to the exotic tastes of the wealthy and powerful. Objects more homely and more domestic than gold and silver or ivory and alabaster would, from the most pessimistic point of view, provide the bonds that would tie up the national imagination. "We are not prepared to allow that wealth is more valued in America than elsewhere," wrote the editor of *The American Whig Review*, "but in other countries the successful pursuit of it is necessarily confined to a few, while here it is open to all." No American was contented to be poor, he continued, and without any established limits there could be "no condition of hopes realized," or, in other words, "of contentment." This desire for physical possessions was "good and hopeful to the interests of the race, but destructive to the happiness and dangerous to the virtue of the generation exposed to it."[6] It was precisely that paradox, then, involving personal ambition and national advancement, which worried the Revolutionary generation.

The search for material happiness and comfort was not confined, during the first half of the nineteenth century, to the mere purchase of objects. It could be demonstrated, among other places, in the how-to-do-it literature which poured from American printers—the advice and directions given for homebuilding and house improvements, for furniture and clothesmaking, for cooking and machinebuilding.[7] Materialism, in the midnineteenth century, was tied both to production *and* consumption. If a man desired what he could not yet afford to buy, he could either enlarge his fortune or apply his own talents to imitation. Instead of going to expensive monopolists of skills and products, expensive artisans and urban merchants, the individual consumer still had the capacity to create some of his own luxuries.

It is, I believe, the declining ability to manufacture one's own material environment that precipitates a growth of interest in the purchasing act. It first develops slowly in the last half of the nineteenth century and then flowers in the first few decades of the twentieth century. We are all familiar with the decline of household production, one of the results of the continuing process of industrialization, technological change, and urbanization. We are also familiar with the growing incapacity of consumers to match, in precision, variety, attractiveness, and especially cost, the profusion of objects produced by American manufacturers, from clothing and furniture to food and drink. To enhance consumer appeal, dramatic improvements in advertising and distribution capacities were made in the late nineteenth century. Newspaper and magazine advertisements grew in number and improved in quality, thanks to printing innovations and the new halftone engraving processes. Mail-order catalogs and a revolution in poster art enhanced display methods and increased public sensitivity to the attractions and variety of purchasable objects. And the enlargement and improvement of retailing operations—in large part through the construction of huge department stores, which achieved their modern physical forms in the twenty-five years preceding World War I—added to retailing temptations, as did the refinements of window dressing.[8]

Even before the Civil War, American magazines were peppered with stories about and comments upon the new retailing. *Harper's Weekly* in 1857 ran one picture of a shopping crowd entitled "The Dry-Goods Epidemic" and spoke of women "with faces like hawks and fingers like claws" surrounding the shopping counters. "The shopping mania," wrote *Harper's*, "ought to have received more attention than it has from the faculty. It is a species of absorbing insanity."[9] In the next several decades it would get more attention. Writing in *The Century* in 1901, Lillie Hamilton French sought to capture the spirit of New York's shopping district below Madison Square:

> There is no one part of the day from eight in the morning until six at night when the stores themselves are not full, and when out of doors you do not have to elbow your way . . . through throngs of people, men, women, and children, old and young, rich and poor . . . The shop-windows, with their elaborate displays, their free exhibitions of the fashionable and the beautiful, are never without their crowds about them.[10]

The problems of consumption, shopping, and materialism were, by the late nineteenth century, increasingly identified with the classes and masses of the great cities, and it was probably inevitable that American novelists of this era, particularly those identified with a concern for real-

"The Dry-Goods Epidemic," an early depiction of the American "shopping" phenomenon. (From Harper's Weekly, 31 Oct. 1857; Smithsonian Institution Neg. No. 81–508.)

ism, should analyze the relationship between consumer desire and personal wealth. They began to chart the rituals of American consumption, the new iconographies of possession and the competition for domestic luxury which lay behind the marketing success of so many American companies. William Dean Howells, the most perceptive of the social realists, observed the crowds of women thronging Washington Street in Boston, "intent upon spending the money of their natural protectors." Unconsciously echoing the imagery of *Harper's* a generation before, he caught the fierce intensity of the female shoppers, clasping parcels to their hearts and coordinating the immense task of moving from store to store in search of bargains which they could tote back to their suburban homes.[11]

Howells's consumers, his admirers of fine things, were caught by traditional problems. An ever-broader slice of the population, they were still dazzled by visions of furniture, carpets, chandeliers, and bric-a-brac. And they fell prey to the moral dilemmas and unsatisfied desires that inevitably ensued.

Howells's greatest novel of social climbing and material desire, *The Rise of Silas Lapham*, focused upon this traditional dilemma by examining an ancient theme—the problems of the newly rich. The great symbolic act which informs the novel—Lapham's commissioning of a new house in Boston's Back Bay—is a translation of an immemorial gesture, indulged in by the new noblemen of Elizabethan England, the tidewater planters of the eighteenth-century Chesapeake, and the industrial millionaires of New York's Fifth Avenue. Lapham's own confrontation with social pretension and commercial dishonesty, the contact between parvenu money and genteel poverty, the personal tensions of urban living, all were explored by Howells with particular brilliance.[12] But the experience of consumption, the lure of material objects, did not absorb him as much as the comedy of manners provoked by the new scale of income.

There is, to be sure, a new level of specificity in Howells's description of the urban landscape and the interiors of middle-class housing and urban flats. The relationships drawn between personality and setting, between temperament and physical object, reached their apogee in Howells's famous descriptions. But the objects are generically classified and form the staple clutter, with some exotic twists, with which the middle classes had surrounded themselves from the midcentury on. In *A Hazard of New Fortunes* the Basil Marches, searching for a furnished flat, encounter the bewildering forest of objects that make up the middle-class interior, and Howells enjoys recounting the details. Mrs. Grosvenor Green's rooms, for example, fairly swarmed with "gimcracks":

The front of the upright piano had what March called a short-skirted portière on it, and the top was covered with vases, with dragon candlesticks, and with Jap fans, which also expanded themselves bat-wise on the walls between the etchings and the water-colors. The floors were covered with filling, and then rugs, and then skins; the easy-chairs all had tidies, Armenian and Turkish and Persian; the lounges and sofas had embroidered cushions hidden under tidies. The radiator was concealed by a Jap screen, and over the top of this some Arab scarfs were flung. There was a superabundance of clocks. China pugs guarded the hearth; a brass sunflower smiled from the top of either andiron, and a brass peacock spread its tail before them inside a high filigree fender; on one side was a coal hod in repoussé brass, and on the other a wrought-iron wood basket. Some red Japanese bird-kites were tucked about in the necks of spelter vases, a crimson Jap umbrella hung opened beneath the chandelier, and each globe had a shade of yellow silk.[13]

A furniture warehouse, in short, the loot from the bazaars of the world, its chief defect its clutter and eclecticism, its willful insistence upon disguise, upon torturing the material or the function or the appliance into something it was not meant to be. It was easy to project from these leavings the persona of their owner. The foibles of fashion formed a vulnerable and venerated target. Howells was cataloging objects with brilliance and effecting a new social translation from an old mode of expression, but he was not transforming his characters fundamentally. Beyond the detail, this was a category of consumer which epitomized bourgeois acquisitiveness and, as such, had been around for quite a few decades.

There were some similarities of focus to be found in a contemporary of Howells's whose works are also filled with the physical details of America's Victorian age; but there is also now a new concern with the effects of consumership and mass appetite upon the life and ambitions of the working class. Theodore Dreiser's picture of the city, his evocative theatrical sets, fit a very old image of glitter and corruption, an image that moralizers of the seventeenth and eighteenth centuries would have recognized. But Dreiser's characterization added, through its drama and absorption with the effects of the material environment upon character, the manipulation of desire and appetite and portrayed a new setting for the terrible implications of American materialism. The corpus of Dreiser criticism and commentary forms a literature much larger than Dreiser's own output, and I am touching only lightly upon one of many directions in Dreiser's thinking. But many critics have noted his absorption with settings designed to enhance the urbanite's role as consumer—the theater, the hotel, the restaurant, and, above all, the department store. *Sister Carrie*, the novel which presented the drama of urban desire most evocatively, employs all these scenes and expresses Dreiser's fascination with the signalling devices, particularly the clothing, by which nineteenth-century Americans assumed their specified roles in the social drama. Within days of her hegira to Chicago, Carrie Meeber, as part of her search for work, finds herself in one of Chicago's new department stores, The Fair, which touches off Dreiser's brief disquisition on the history of this commercial enterprise. Carrie's encounter with the retail goods stands, in miniature, for her encounter with the great promise of prosperity.

> Each separate counter was a show-place of dazzling interest and attraction. She could not help feeling the claim of each trinket and valuable upon her personally, and yet she did not stop. There was nothing there which she could not have used—nothing which she did not long to own. The dainty slippers and stockings, the delicately frilled skirts and petticoats, the laces, ribbons, hair-combs, purses, all touched her with individual desire, and she felt keenly the fact that not any of these things were in the range of her purchase.

Carrie's appetite and her sense of diminution are increased by the sight of other shoppers, "the fine ladies who elbowed and ignored her, brushing past in utter disregard of her presence, themselves eagerly enlisted in the materials which the store contained." The very shopgirls, pretty and well dressed, make her feel uncomfortable.[14]

The effort to make herself a respectable consumer, a shopper who not only looks but buys, forms one of Carrie's buried motives for leaving the dreary flat inhabited by her sister and brother-in-law. And, within a few chapters, beginning to respond to the wiles of Drouet, a drummer, Carrie is able to return to The Fair, to look at jackets. Dreiser entitled this chapter "The Lure of The Material: Beauty Speaks for Itself," and it is filled with images of Carrie's desire for finery, for shoes, stockings, a skirt and jacket. In The Fair, with Drouet's money, she indulges in one of the new consumer passions. "There is nothing in this world more delightful than that middle state in which we mentally balance at times, possessed of the means, lured by desire, and yet deterred by conscience or want of decision." Here, in this exploration of consumer sensibility, Dreiser has Carrie wander around the store, examining pieces of finery. "Her woman's heart was warm with desire for them. How would she look in this, how charming that would make her!" She moves from the corset department to jewelry, and Dreiser alludes to the earrings, bracelets, chains, and pins that she lovingly encounters. "What would she not have given if she could have had them all!"[15] Accompanying her in the store, advising her on purchases, Drouet employs the rich mercantile atmosphere as part of his courtship. The department store becomes the garden of desire, as suitable for American lovers at the turn of the century as a formal bower might have served couples centuries before. In describing the shopping rituals, the uncharted course of Carrie's wants, and the use of merchandise as a lure to win over her hesitations and suspicions, Dreiser has seized upon a central emblem for the new urban society. And throughout the book he uses the presence of consumer desire as a fever chart to indicate the health of his protagonists.

Thus Hurstwood's dissolution in New York, after the saloon owner abandons his family and flees with the stolen money, is defined in part by his demeaned consumer status, his dependence upon the sufferance of tradesmen, and his need to bargain and learn comparative shopping. He now grows aware of the price of meat, butter, and coal, searching the neighborhood for bargains or trying to do without the little luxuries. Carrie, having learned the pleasures of shopping without fear, reacts with bitterness. Such "miserable details ate the heart out of Carrie. They blackened her days and grieved her soul." Dreiser catalogs the minutiae. "Hurstwood bought the flour—which all grocers sold in $3\frac{1}{2}$ pound packages—for thirteen cents and paid fifteen cents for a half pound of liver and bacon." Carrie, meanwhile, quarreling about debts with the milkman and the coal seller, goes for a drive, her eye "once more taken by the show of wealth—the elaborate costumes, elegant harnesses, spirited horses, and above all, the beauty. Once more the plague of poverty galled her, but now she forgot in a measure her own troubles so far as to forget Hurstwood."[16]

In detail, in drama, in its extraordinary setting of commercial and domestic scenes, Dreiser's novel broke new paths. But its ultimate meaning was, in many ways, highly traditional. At the end Carrie's search for material happiness contains a ring of doom. Surrounded by "her gowns and carriage, her furniture and bank account," by friends and the applause of her audiences, she remains unsatisfied. "Amid the tinsel and shine of her state walked Carrie, unhappy."[17] The objects surrounding her were reduced to false witness in the moralizing last vignette. Dreiser's challenge to conventional morality was his failure to punish Carrie for her fall from grace, leaving her with a successful stage career stretching ahead while Hurstwood commits suicide. But the challenge was not fundamental. In denying Carrie happiness Dreiser simply extended an older, pietistic notion of the limitations of material success when divorced from spiritual progress. And his portrayal of the seductions of metropolitan life, along with his homilies on the happy family, could have fit comfortably within a whole series of more conventional sermons on the dangers of unchecked appetite.

Traditionalism also pervades human-object relationships in Dreiser's great trilogy on Frank Cowperwood. Cowperwood employs his wealth, in the second and most brilliant volume, *The Titan*, to establish himself as an economic and social force in his adopted city of Chicago. He pursues the usual millionaire's course of building a great house, collecting art, covering his wife with jewels, and generally maintaining the entourage of a grandee. Cowperwood, however, is neither dazzled nor seduced by his wealth. His object remains power, and for him power is a headier experience than mere financial expenditure. The corruptions he dispenses only emphasize his own transcendence of ordinary constraints and appetites, the superman-tycoon who shrugs off wounds which would fell ordinary mortals. Cowperwood in a sense is one of Tocqueville's aristocrats, self-made, but tolerating material losses with equanimity, made more determined than ever by setbacks and opposition. His conquest of society, incomplete and unsuccessful, nonetheless rests on a masterful self-discipline which stands impervious to the tinsel and glitter that seduce weaker spirits. The luxury surrounding the great businessman forms a series of props to his public presence, and, with the exception of his art collection, arouses none of the passion which it did in the heart of an arriviste like Carrie.[18]

The work of Howells and Dreiser, as the output of individual novelists, can hardly be termed representative, nor can it be fastened securely within the terms of an argument about consumer desire. But it testifies to the fact that well before World War I the buying drama had begun to serve as a symbol for modernity, and the buying experience had become a ritual worthy of examination, a metaphor for national mobility, social climbing, economic competition, and moral deterioration. On the whole the buying drama was still confined to women; its objects were generically described; its corruptions came principally through luxurious specialty items or through the transplanted and temporarily dispossessed, who would endure any privation to obtain a larger stock of the world's goods. But a change in sentiment was coming, a change which had its origins in the pre-World War I years but which intensified in the 1920s. American novelists in the twenties would also pay attention to the great American consumer, but there would be a shift of focus, a transformed manner of handling the relationship between people and mass-produced objects, a broader sector of the population to describe, an increased specificity, and a widening of critical attention to the buying act. The sensuality which Howells and to a much larger extent Dreiser located in the act of material possession now becomes a mass experience not confined to the stores or expensive merchandise. We enter, through the novels, a new world of merchandising, one closer to our own day. But before moving to consider these novels it would be useful, I believe, to review some aspects of the merchandising revolution in progress on the eve of the war, and intensifying during the following two decades. Each aspect I describe is merely a summary of a process which has received, or is capable of receiving, far more extensive and detailed treatment. But together they form what must be seen as a historic moment in the history of American selling: a moment which penetrates the world of American fiction, as novelists sense the struggle of the merchandisers to serve a new social type, and the needs people place upon objects as part of their social orientation.

One element of the merchandising revolution lay in the physical reconstruction of American department and specialty stores. Shopping architecture in the major cities was transformed as American retailers tried to incorporate some of the possibilities of modernism, as they had been evidenced at the great 1925 Exposition des Arts Decoratifs in Paris. Store owners, confronting narrow profit margins, increasing downtown congestion, the beginnings of the suburban exodus, and the challenge of chain stores and mail-order houses, began to mount a new and more aggressive campaign to capture consumer loyalties. "It is only in comparatively recent years," R. W. Sexton wrote in 1928, "that merchants have been brought to realize that art has a selling value. The keen competition amongst stores and shops in the congested shopping districts of the larger cities has already been responsible for the maturing of this appreciation of the commercial value of beautiful things." In a chapter on "Stores and Shops," Sexton, a student of contemporary American commercial and residential building, argued that well-designed shops could differentiate themselves from one another through their displays, luring the "prospective customer in the expectation of finding there the unusual, the unique, things which they may not find other places."[19] It was once believed, Sexton argued, that customers did not wish to be distracted from serious buying missions when entering a store. But driven by keener competition, shop owners now realized how crucial was eye appeal to retailing success. The new art of commercial display, John Taylor Boyd asserted in the *Architectural Record*, "aims less at making an exhibition of individual wares than it does at portraying those ideas of luxury, fashion, or style now so important in retail trade, as well as at impressing the public with the mastery on the part of the shop of fashion and style."[20] In place of merely showing goods, stores were displaying tableaus and pictures suggesting locale, mood, or historic incident. Captivated by such pictures, customers stopped, to purchase their ideals of fantasy as well as the specific commodity. Contemporary stores could not wait until customers were in absolute need of merchandise, Frederick J. Kiesler warned. "You must create demand."[21]

The redesign of store facades and interiors focused attention on the pervasive field of action available to the intelligent seller, on the lure of well-displayed objects and their aesthetic properties. Buying was now more than merely a survey of luxurious objects; it was a total experience in which the retailer became a subtle adviser on personal taste, a joiner who fit individual temperament with proper merchandise. The elegant store was less a bazaar and more of a fitting-room. The shop itself, rather than the array of items it displayed, became a muscular advocate for the buying drama.

Simultaneously there were other developments which demonstrated increasing sensitivity to the power of objects in exciting emotional loyalty and which increased awareness of their physical properties. One was a series of improvements in color printing and engraving, improvements which permitted large, full-page, full-color magazine advertisements to be far more faithful to the colors and textures of objects—particularly foods, textile, and automobiles—than ever before. Effective halftone advertising techniques were now more than a generation old, but problems of color reproduction had only recently been solved. This new control over visual advertisement gave manufacturers and retailers a further weapon in their struggle for popular interest.[22]

A third factor in expanding the issue of consumer consciousness could be found in the rationalization of advertising methods developed during the 1920s, stimulated in part by the experiences of World War I and by the more systematic application of experimental psychology to problems of motivating and channeling desire. This rationalization also owed much to the increasing self-confidence and articulateness of the advertising industry itself. Again the roots were prewar, but the science of selling was given, during the 1920s, special deference and attention. "Perhaps the most widely discussed subject in business today is the art of selling," Floyd Parsons announced in the *Saturday Evening Post* in 1920. "Efforts are now being made to develop definite methods for use in the selection of salesmen," he went on. The seller of goods must now be a psychologist, a statistician, a personality analyst, and a performer, in addition to possessing the skills of a merchant.[23] Book after book was written to describe the talents appropriate to salesmanship and to celebrate its social value. Bruce Barton's *The Man Nobody Knows* was merely the most famous example of a huge literature which justified the costs and energies consumed by selling and its advertising components. The rise of the public-relations counsel and the selling of character and personality were simple extensions of the method, once the parameters had been fixed.

Enough work has been done on the modern history of American advertising for this subject to be introduced without much further detail. The same is not true, however, of a fourth element enhancing the social role of consumer goods: the expansion of the American film industry. Profiting from the collapse of European competitors in the wake of World War I,

American films dominated not only in this country but throughout the world market, establishing an international reputation for their stars, directors, and producers. Film's influence on consumer products, however subtle and complex, was probably almost as important as its provision of a new set of celebrities, for the image on the movie screen inevitably focused attention on the objects which formed part of its decor. The lingering close-ups, the use of music to emphasize mood, the employment of objects and sets as significant aspects of plot and character development, all emphasized the sensuous properties of what might have been seen, more casually, as mundane artifacts, hardly deserving of sustained attention. Because of the variety in film subject matter, the objects redeemed by the camera ran the gamut from expensive playthings, traditional objects of luxury, to the ordinary appliances of daily life. The presence of a certain style of clothing, a set of furniture, an interior décor, in a major film, could touch off considerable public demand. Films provided an underpinning for American faddishness, running all the way from hair-dos and voice inflections to the design of offices and domestic interiors.[24]

Some of the object focus in film resulted from the association with stars, whose own consumption patterns received respectful and detailed attention from the news magazines, newspapers, and fan journals. The Hollywood style, a new blend of luxury and informality, projecting some of the qualities of southern California, set a consumption pattern that exerted considerable mass appeal.[25] Although the houses, the swimming pools, the extravagant art and furniture were clearly beyond the reach of most consumers, the cinema aristocracy inspired a good deal of less expensive imitation. In some ways this was a modern version of a traditional mode of style setting, except that the pacesetters were now the creatures of a mass-culture industry rather than socialites, military figures, wealthy businessmen, or any of the previous groups of fashion leaders. But the capacity of the camera to exaggerate material luxury, the exoticism of Hollywood's location, the size of the mass audience attending films, and the continual publicity given the stars, all created an unprecedented intensity and sharpness of focus. It is possible, occasionally, to document film's influence on fads or fashions. More difficult—and probably more important—are the undocumentable aspects of film's influence: the custom it inspired of examining the surfaces, shapes, and dimensions of objects with new interest as a result of their appearance on the screen; the stimulation of consumer consciousness because of the screen's capacity for memorable exaggeration. And all this was enhanced, of course, by the gorgeousness of the movie palaces: their opulence underlined the luxurious dreams played on the screen.

The new stores, the improvements in photography, the self-consciousness of the advertising industry, and the role of motion pictures were supplemented by other trends emphasizing the capacity of mass-produced objects to shape daily life. One was the advent of the industrial-design profession, a group of men that included Norman Bel Geddes, Raymond Loewy, Henry Dreyfuss, Gilbert Rohde, Lee Simonson, and Walter Dorwin Teague. These men often had backgrounds in theater and set design and worked on products that would not only function effectively but create more compelling, saleable images.[26] They were client oriented; several had already been employed to redecorate department stores and create retail liveries for store uniforms, delivery vans, wrapping papers, and so on. Their purpose was to demonstrate, to manufacturers and retailers, that attention to physical appearance produced larger profits. Loewy scored an early success by redesigning the Lucky Strike package and convincing hard-bitten George Washington Hill that he was worth his designer fee.

The industrial designers also claimed more disinterested objectives, seeking to develop art forms within machine-made objects that did not challenge the logic of their origins. Dispensing, as they did, with traditional decorative devices, sympathetic with contemporary European design, aware of the progress of the International School, the industrial designers confronted a characteristic dilemma: how to increase the allure of their products without relying upon the sentimental details and frills that had gained them customers at an earlier date. A certain amount of public education was necessary for potential audiences to appreciate the new uncluttered surfaces, the smooth curves, the dramatic lettering, the carefully planned knobs and controls in the redesigned refrigerators, stoves, alarm clocks, lighters, vacuum cleaners, hot-water heaters, and radios that flew off their drawing boards. But aided by the new photography, by the crusading of art publicists eager to legitimize modern design, by movies, by the railroad companies that popularized streamlined design in their search for more customers and were major patrons of Loewy and Dreyfuss, the industrial designers made considerable progress.

They were helped, also, by the great international expositions of the 1930s, in New York, San Francisco, and Chicago, where their work was displayed quite effectively.[27] Norman Bel Geddes, designer of the G.M. Futurama, and Walter Dorwin Teague, employed by Eastman Kodak, U.S. Steel, Du Pont, and Con Edison, were among the most visible designers of their generation. Surrounded with drama, greeted with excited enthusiasm by the press, their newly shaped objects achieved special immediacy and identity and seemed more intrinsically interesting than their predecessors. The movement by institutions like the Museum of Modern Art to collect well-designed objects of daily use added force to the designers' insistence that they were seeking artistic glory as well as increased sales. And the possibility of purchasing a mass-produced masterpiece added a certain piquancy to the consumer's activities. In sum, the contribution of industrial design provided, at a minimum, reinforcement to those seeking to increase the emotional energy invested in the adoration of mass-produced objects. And the designers managed this without associating such self-consciousness with dandyism or affectation, its standard earlier associations.

The achievement of the designers was complemented by another feature of the interwar years which has received extensive commentary in another context, and that was the popularity of the model change, particularly for automobiles.[28] It fitted easily into the heightened consciousness of object form that I have been trying to outline. Now it was not only the

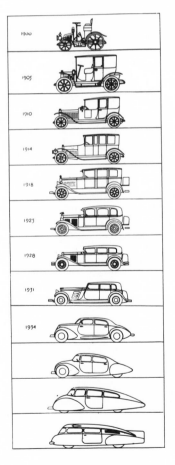

Raymond Loewy's "Evolution Chart of Automobiles," 1933. (Courtesy of Raymond Loewy.)

particular model that was important, but its vintage as well; the consumer was moving through time as well as space. How extensively and frequently manufacturers of consumer staples put model changes into effect varied, of course, according to the object, the state of technology, and the intervention of industrial designers. But there begins to be, in the 1920s, far less permanence in the appearance of many consumer products. This was one of the features that stimulated retailers to get involved with store renovation and to reconsider the appearance of the buying environment. A concern with up-to-the-minute fashion had once been reserved almost entirely to women's clothing; now it was extended to many other things. The consumer's notice was caught by continual demands that he move up to a new product, for the old model was deprived, through advertisement, of those magical intangibles which the copywriters fixed only on the newer model. It was also possible—in the case of automobiles, radios, phonographs, and refrigerators—to talk about technical changes which promised improved performance. The model change thus joins a lengthy list of influences stimulating sensitivity to the aesthetics of consumer goods.

One final element must, at least, be considered. It was not novel to the postwar period but rather was the final phase of a competitive era about to end. Combined with the other factors it was important. And this element was the number of brands available within any single object category. The coexistence of national advertising and a tremendous set of brand choices made the advertising pages of magazines take on almost encyclopedic detail. Looking back at the 1920s from the present, the merchandising world looks bewildering. In the midtwenties, May Hoffman, director of research for a major advertising agency, examined the buying habits of small-town women, and in particular examined their brand preferences. The tabulated results are, in some ways, less interesting than the raw data. Some 210 women, for example, reported having purchased pianos in a town located near Kansas City, Missouri. But among the 210 the most popular brand, Gulbransen, had only ten buyers; Chickering, Kimball, Vose, and Steinway followed with six each. All the rest sold three or less. Among the 210 buyers, 101 separate brands were represented, most of which are no longer with us. For phonographs, although Victor was heavily dominant, there were twenty-six different brands represented, in addition to thirty manufacturers of radios, fifty different kinds of washing machines, and twenty-five different automobile makes, including the Essex, Oakland, Hupmobile, Franklin, Hudson, Maxwell, Packard, and Elcar.[29] Consumers confronted an extraordinary range of choices, and the competing manufacturers were forced into marketing strategies as ingenious and aggressive as the hard-pressed department stores. Like the model changes, this variety of brands stimulated interest in the object's symbolic properties. In the end it seemed to make the act of consumption a more significant statement of choice and preference and to increase the curiosity of friends and neighbors about the logic and meaning of specific buying patterns.

The range of trends, experiences, and innovations that I have just summarized, accompanied, I believe, a newer and sharper fictional exploration of the role and character of American consumption and helped stimulate a consideration of the function consumer goods performed in the imaginative lives and daily activities of many Americans. The shift had begun even before the war. But in, the 1920s a larger discussion developed, accompanied by different emphases upon the consumer audience, its dependence upon material artifacts for identity as well as for identification, and the iconographic significance of brands and products. The wave of cultural criticism that succeeded the war, involving figures like Van Wyck Brooks, Harold Stearns, H. L. Mencken, George Jean Nathan, and Lewis Mumford, set the scene. In their *American Credo* of 1920, Mencken and Nathan redeemed Americans from charges of avariciousness and miserliness, only to substitute in their place a childlike insistence upon acquisition. Where an older generation put its money into banks or real property, "the young folks put their inheritance into phonographs, Fords, boiled shirts, yellow shoes, cuckoo clocks, lithographs of the current mountebanks, oil stock, automatic pianos," and in general into anything that could glitter and distract them. "Whatever is shiny gets their dollars.... They are, so to speak, constantly on a bust, their eyes alert for chances to get rid of their small change."[30] Lewis Mumford found the tension between the desire for acquisition and collection, and the limited income of most workers, to provide the source for urban amusements; a temporary resolution came through the invention of the five-and-dime store. "People who do not know how to spend their time must take what satisfaction they can in spending their money. That is why, although the five and ten cent store is perhaps mainly an institution for the proletariat, the habits and dispositions it encourages are universal." The chief amusement of Atlantic City, Mumford went on, consisted not of its beach, or ocean bathing, but visiting the shops that lined the boardwalk. Consumption, in the eyes of these critics, had become a bribe; the "coercive repression of an impersonal, mechanical technique," Mumford went on, with his eyes on the larger industrial system, "was compensated by the pervasive will-to-power—or at least will-to-comfort—of commercialism."[31]

The corruptions that the eighteenth-century political scientists had feared, now flowered. The use of objects as distraction; their standardized plenitude (another example of the passion for conformity that seemed to have become a permanent presence in American civilization); and the substitution of goods and mechanical exercises for direct experience and spiritual values made the act of consumption take on larger overtones than it had for the prewar generation. To describe it became an act of criticism, criticism not of individual protagonists but of a larger system. Heavy reliance upon objects for satisfaction of psychological, sexual, or social needs became a characteristic theme for several novelists, and the relationship between Americans and their objects—their methods of buying, their systems of classification, their patterns of use—took on broader significance. Through further specification of product, dissection of the buying process, exploration of the psychology of the consumer, and, above all, direct connections between consumption and identity, several novelists of the decade created a portrait of the American buyer, and a revelation of changes in national character that remains evocative.

Thus one can turn to a writer who is not normally associated with the economic novel, F. Scott Fitzgerald. Fitzgerald's work is revealing, for purposes of this discussion, not because of the centrality of his interests in consumption, but because of certain changes of emphasis. Focusing upon the wealthy, upon consumers of luxury goods, Fitzgerald no longer presented generic categories, like Howells and Dreiser. He enjoyed the specific details, the brand names, the qualities of the object. It was part of the persuasive power that the Fitzgerald settings achieved. It may not seem much in 1945, John O'Hara wrote in his introduction to a Fitzgerald anthology, "but twenty-five years ago it was delightful to find a writer who would come right out and say Locomobile instead of high-powered motor car." Fitzgerald, insisted O'Hara, always knew his situations; unlike other writers he wasn't dependent upon the local branch library. "The reader usually knew, without stopping to think much about it, that if a family owned a Franklin it was because they didn't feel they could afford a Pierce-Arrow."[32] O'Hara himself, in *Appointment in Samarra*, was to make much of the subtle distinctions among automobile makes, and he valued Fitzgerald's specificity as part of a pattern of description crucial for determining questions of taste, temperament, and class.[33]

Fitzgerald was concerned, moreover, with the consumption patterns of men as well as women, another new development for writers of the decade. The merchandising revolution of the postwar years involved both sexes; Fitzgerald's greatest consumer was, of course, Jay Gatsby. The pleasure domes constructed by the seekers for power included, among their brilliant attractions, a profusion of more ordinary items. Gatsby's shirts, for example, imported in shimmering colors and fabrics from England, achieve, through number and splendor, the status of a treasure house. Filling the "two hulking patent cabinets," "piled like bricks in stacks a dozen high," the shirts are displayed for Daisy, thrown on a table. The "soft rich heap mounted higher—shirts with stripes and scrolls and plaids in coral and apple-green and lavender and faint orange, with monograms of Indian blue," so beautiful that Daisy sobs in wonder.[34] The sadness and awe at Gatsby's innocent profusion catches, in miniature, the tragedy of denied fulfillment that Fitzgerald lifts to mythic proportion. And in another vision of the soured promise of America, the 1922 short story "May Day," Fitzgerald uses the image of piled-up shirts to project a contrast between two Yale classmates, Gordon Sterrett, who will commit suicide by the story's end, and Philip Dean, selfish, wealthy, and self-confident.[35] "May Day" is filled with references to commercial New York, discussions of haberdashery, reconstructions of shopping drama. "Fifth Avenue and Forty-fourth Street swarmed with the noon crowd. The wealthy, happy sun glittered in transient gold through the thick windows of the smart shops, lighting upon mesh bags and purses and strings of pearls in gray velvet cases; upon gaudy feather fans of many colors; upon the laces and silks of expensive dresses. . . . " The theme of unrequited love for earthly treasure is reconstructed by Fitzgerald to suit the urban crowd, working girls who, while they inspect wedding rings and platinum wrist watches, digest "the sandwiches and sundaes they had eaten for lunch."[36]

The sensual appeal of stores and the central modern experience of shopping are themes carried out in several Fitzgerald works, *Tender is the Night* particularly; according to some critics, they help give this novel much of its bite and irony. In one memorable and frequently quoted passage, Fitzgerald placed Nicole, daughter of a wealthy North Shore Chicago family, at the apex of a system that ran around the world:

> For her sake trains began their run in Chicago and traversed the round belly of the continent to California; chicle factories fumed and link belts grew link by link . . . men mixed toothpaste in vats and drew mouthwash out of copper hogsheads; girls canned tomatoes . . . half-breed Indians toiled on Brazilian coffee plantations and dreamers were muscled out of patent rights in new tractors—these were some of the people who gave a tithe to Nicole, and as the whole system swayed and thundered onward it lent a feverish bloom to such processes of hers as wholesale buying like the flush of a fireman's face holding his post before a spreading blaze. She illustrated very simple principles, containing in herself her own doom . . . .[37]

Shopping is no longer, as it had been for Dreiser and his generation, the symbol of individual temptation or even personal exploitation. It has become an image for a planetary economic system, judged and found guilty of brutality, inequity, and even, in Fitzgerald's terms, of madness, doomed, in the long run, but meanwhile inspired to frantic activity. Consuming is one of the few art forms left to individuals; their expression of purpose and self must be in terms of what they can buy, for there is no other standard of value. When added to Fitzgerald's concern with enumeration and specification and placed within the richly populated world of his stories and novels, the social indictment exerts enormous power. The objects are no longer adjuncts; they dominate, dangling below them the personality of the protagonist. The American system of consumption has become a drama in which an old, sacred dream about the redemptive power of a landscape has been translated into a secular conviction about the saving power of merchandise.

But Fitzgerald's protagonists still represent a special section of the consumer world, the very rich. This was the setting he knew best, and in the 1920s the drama of buying shifted to a much larger public and to a far wider range of products, as my earlier survey attempted to indicate. And here, the most detailed, comprehensive, and emphatic portraits were drawn by the novelist of the American middle class, whose books from the twenties represent a commentary on national buying habits: Sinclair Lewis. Lewis was early attracted to the setting of the middle-class commercial world; one of his first novels, *The Job*, published in 1917, concerned itself with the new career woman and the state of manners and morals of metropolitan business. But it was in the twenties that his most popular books were published. In three of them, *Main Street*, *Babbitt*, and *Dodsworth*, all in print by 1929, Lewis examined buying, selling, and advertising in greatest detail.

In his brief preface to *Main Street*, Lewis stated explicitly the ironic relationship between the town culture he was examining and the course of human history. "Main Street is the climax of civilization. That this Ford car might stand in front of the Bon Ton Store, Hannibal invaded Rome and Erasmus wrote in Oxford cloisters." [38] The theme that Fitzgerald would imply in *Tender is the Night* was here stated more baldly. Carol Kennicott, absorbed by the task of bringing Gopher Prairie into the mainstream of civilization, is appalled not simply by the library, the schools, the theater, and other traditional hallmarks of the higher life, but by commerce, by buying opportunity, by merchandising displays. Her first tour of the town focuses upon the dull, sometimes dirty, invariably repellent atmosphere of the stores and shops, relieved only by the modern plate-glass windows of Harry Haydock's Bon Ton Store. Lewis's target in this architectural tour is apparently the rampant individualism of American architecture and town planning, the failure of buildings to harmonize with one another, the clutter of light poles and telegraph lines, gasoline pumps, and litter. "Each man had built with the most valiant disregard of all the others." [39] But this variety did not increase choice. Only in the metropolis—Minneapolis or Chicago—could the intelligent and venturesome consumer find goods to satisfy her heart. Small-town America, the target of so much cultural criticism in the early twenties for its creative repressiveness, its tyranny of opinion, its smugness, its xenophobia, its intolerance, provided too limited a field for Lewis to exercise his full descriptive skills, and, several years later, in *Babbitt*, he was able to give play to his concerns about consumer culture.

Babbitt remains perhaps the most detailed portrait of American consumption in our fiction. As Mark Schorer has pointed out, it is hardly a novel at all. As "the major documentation in literature of American business culture," the book consists of a series of "set pieces, each with its own topic, and all together giving us a punctilious analysis of the sociology of American commercial culture." [40] Individual chapters address domestic manners, marriage, Pullman Car culture, leisure time, conventions, and other features of bourgeois life. *Babbitt* consists of an endless set of inventories—inventories of clothing, of furniture, of automobile fixtures, of bathrooms and barbershops and haberdashers. Lewis is at pains to outline the symbolic system which surround the objects. Eyeglasses, shoes, and suits present the outer man, and automobiles fix his rank in the social world. In the city of Zenith, wrote Lewis,

> a family's motor indicated its social rank as precisely as the grades of the peerage determined the rank of an English family—indeed, more precisely... There was no court to decide whether the second son of a Pierce Arrow limousine should go in to dinner before the first son of a Buick roadster, but of their respective social importance there was no doubt; and where Babbitt as a boy had aspired to the presidency, his son Ted aspired to a Packard twin-six and an established position in the motored gentry. [41]

The system of correspondences fixed by Lewis attracted his scorn precisely because it had become a system—regulated and fixed by external authority. "These standard advertised wares—toothpastes, socks, tires, cameras, instantaneous hot-water heaters—were his [Babbitt's] symbols and proofs of excellence; at first the signs, then the substitutes, for joy and passion and wisdom." [42] Tocqueville's prophecy could not have been more precisely summarized. The ownership of objects has become the chief goal of living, and self-understanding is possible only through contemplation of material possessions.

The standardized consumer goods that American industry labored to produce aroused in Lewis warring tendencies. On the one hand they symbolized reduced decision-making power, the buyer becoming simply the passive object of manipulators who were presenting him with objects which so closely resembled one another that there was little exercise of personality in the act of purchase. And since so much was now invested in the kinds of objects adorning the person, this loss of power threatened individuality. On the other hand, Lewis admired the predictability, the comfort, the convenience and economy that standardization brought. In a debate taking place early in the book, Seneca Doane, Zenith's radical lawyer, admits to an affection for standardization even while he opposes conformity. The Ingersoll watch and the Ford car were, after all, reliable. "I know precisely what I'm getting, and that leaves me more time and energy to be individual in... There's no other country in the world that has such pleasant houses. And I don't care if they *are* standardized."[43]

The relationship between the proliferation of comforts and the declining standards of excellence absorbed Lewis throughout the decade. There were many streams feeding this interest, going back to earlier generations of American artists, among them a wilderness urge which abhorred the soft life of the cities and found redemption only in the solitary confrontation with nature. But more often than not it was the metropolis rather than the wilderness which Lewis employed as his counterweight, a center of sophistication and cosmopolitanism in which broader and more sophisticated displays of consumer goods testified to a toleration for heterodox opinions. Merchandising, then, occupied a curiously ambiguous place in Lewis's scheme of things: on the one hand, testimony to the commercialization of American culture, the triumph of mass-produced objects over personality; and on the other, evidence of taste, culture, artistic accomplishment, and sophistication. Buying could be either an act of subservience to manufacturers and advertisers or a demonstration of individuality.

This bifurcation becomes most apparent in the third volume of what could be labelled a trilogy of consumer drama, *Dodsworth*, published in 1929. In this study of an American automobile manufacturer who travels to Europe in an effort to keep his wife and come to terms with the world of culture, Sinclair Lewis confronts most directly the problem of maintaining identity in a world of objects, only this time the objects include both the mass-produced goods of the New World and the luxury items of the Old. The precipitating event in Dodsworth's travels—the sale of his Revelation Motor Company to the giant Unit Automotive Company, with its millions of dollars of capital—was a symbol of the submergence of individual creations within much larger systems. Selling his company, Dodsworth tries to hide from himself the notion that the U.A.C., "with their mass production would cheapen and ruin the Revelation and turn his thunderbolt into a standardized cigar-lighter."[44]

The problem of standardization of goods recurs again and again. In England, Dodsworth is astonished to discover that older societies, spurning American methods, are still capable of producing impressive objects. Strolling down St. James's, he finds a window full of modern weapons. "He had not believed, somehow, that the English would have such beautiful shiny shotguns."[45] European selling he also finds superior to American in subtlety and manner. Commercial images abound in the book, and buying becomes itself a testing experience, a means of measuring the mettle of character. Boasting of bargains has become part of the common parlance of Americans, particularly of Fran Dodsworth, but the bargaining of Europeans takes on style, becomes a ritual which throws into the shadows the crude hustle of the American stores.[46] In the end Dodsworth sees his mission as achieving a union between American mass production and the attention to detail and comfort characterizing European design. Architecture, rather than motorcars, attracts him. "It came to him that now there was but little pioneering in manufacturing motors; that he hadn't much desire to fling out more cars on the packed highways. To create houses, perhaps less Coney Island-like than these—noble houses that would last three hundred years"—that would be interesting.[47] Suburban housing, which could unite the color and irregularity of the Old World with the built-in garages and American incinerators that symbolized modern technology—here was what he was looking for.

So the redemption of the manufacturer lies in reapplying the instinct for workmanship, and the redemption of the buyer lies in selecting goods which permit some kind of personal taste to shape the purchase. Carol Kennicott, the somewhat ineffectual aesthete, and George Babbitt, the mindless purchaser of the advertisers' fantasies, are brought together and reshaped in Samuel Dodsworth, who somehow dreams of supplying this mass market with objects that are not compromised in quality by number or economy. Dodsworth's tours of real-estate developments and foreign automobile factories are in their own way reminiscent of the popularity of the do-it-yourself guides of a century before; the dream of prosperity can be attained, in his case, by active intervention, by a reentry into the manufacturing world, reasserting levels of craftsmanship that are gradually being abandoned. And in so doing the passive, tourist values which Fran Dodsworth exemplifies are repudiated. Submission is the feared luxury, the corrupter, not the provision of better objects for the American consumer.

So brief a summary cannot do justice to Lewis, but it may suggest some problems that novelists of the early twentieth century were beginning to examine in more depth, in particular the relationship between the objects of consumer desire and the creation of personality. It is not simply status that manufactured objects provide for members of the national community, but identity itself. The world of objects is no longer accessory, adjunct, scene-setting, as in Howells, or even Dreiser. And the quest for the object is no longer merely indulgence in sensual luxury and appetite. Objects have so multiplied that the ability to control them, to choose among them, to refuse to be overwhelmed by them, becomes a test of personal strength. The buying ritual is simultaneously an act of initiation, a competitive encounter, an exercise of judgment, and an assertion of individuality. In the 1920s, the flood of goods seemed impossible to dam; corruption could be avoided not by a refusal to participate in the great consumer drama, but by the exercise of choice and the determination of particular relationships between objects and individuals. Just as a system of interchangeable parts was becoming, according to its critics, a system of interchangeable workers who were metaphorically likened to cogs in machines, an image captured at this time by Chaplin, so the American consumer was threatened now with the same kind of interchangeability. Only by an act of will could this consumer avoid the anonymity which mass production encouraged, the infinite replicability of mass-produced furniture, and clothing, and appliances. The efforts of advertising men, photographers, film producers, architects and interior designers, and retailers to link their products with adventure, mystery, and romance, to graft narrative onto inanimate forms, and to develop aesthetic principles by which they could be evaluated, was a major effort of American marketing. One can detect in our fiction a discovery of this process in operation, a concern with explicating and evaluating the modern situation so deeply colored by the consumer sensibility. Novelists witnessed and described the movement of objects from expressions of status to guarantors of identity. In so doing they helped map the ideology of a new American system, a search to individualize rather than standardize, by grafting onto mass production the alluring, psychological qualities that answered private dreams.

Notes

Since this essay's completion in 1978, a number of books and essays have appeared which bear upon its subject. Although these works could not be incorporated into the text, readers might wish to note the following: Gunther Barth, *City People: The Rise of Modern City Culture in Nineteenth-Century America* (New York, 1980); William Leach, *True Love and Perfect Union: The Feminist Reform of Sex and Society* (New York, 1980); Leonard Marcus, *The American Store Window*

(New York and London, 1978); Lary May, *Screening Out the Past: The Birth of Mass Culture and the Motion Picture Industry* (New York, 1980); Jeffrey L. Meikle, *Twentieth Century Limited: Industrial Design in America, 1925–1939* (Philadelphia, 1979); Michael B. Miller, *The Bon Marché: Bourgeois Culture and the Department Store* (Princeton, 1981); and Elizabeth Stillinger, *The Antiquers...Changing Taste in American Antiques, 1850–1930* (New York, 1980).

1. There is a large literature on this theme. Edmund S. Morgan's *The Puritan Dilemma: The Story of John Winthrop* (Boston, 1958), remains an excellent and concise statement. Richard L. Bushman, *From Puritan to Yankee* (Cambridge, Mass., 1967), traces the interaction between the changing socio-economic ethic and the Puritan character. Perry Miller, *The New England Mind, From Colony to Province* (Cambridge, Mass., 1939, 1954), is a classic statement. Bernard Bailyn, "The Apologia of Robert Keayne," *William and Mary Quarterly*, 3d ser. 7 (Oct. 1950): 568–87; and Bernard Bailyn, *The New England Merchants in the Seventeenth Century* (Cambridge, Mass., 1955), are important studies of more specific episodes in the conflict between economic ambition and religious vision. A more recent work, Stephen Foster, *Their Solitary Way: The Puritan Social Ethic in the First Century of Settlement in New England* (New Haven, Conn., 1971), has important material bearing on the issue.

2. Again, there is a very large literature on this subject. The most sophisticated discussion can be found in Gordon S. Wood, *The Creation of the American Republic, 1776–1787* (Chapel Hill, N.C., 1969), particularly chaps. 2, 3, 10, 15. See also Neil Harris, *The Artist in American Society: The Formative Years, 1790–1860* (New York, 1966), chap. 2.

3. Fred Somkin, *Unquiet Eagle: Memory and Desire in the Idea of American Freedom, 1815–1860* (Ithaca, N.Y., 1967). This phrase is the title of Somkin's opening chapter.

4. This relationship, and the power of public opinion, was explored by many foreign visitors during the Jacksonian period. References and summaries are provided in Jane Louise Mesick, *The English Traveller in America, 1785–1835* (New York, 1922); and Max Berger, *The British Traveller in America, 1836–1860* (New York, 1943). An anthology of appropriate materials is *America Through British Eyes*, ed. Allan Nevins (New York, 1948); the second and third sections are most relevant here.

5. Alexis de Tocqueville, *Democracy in America*, ed. Phillips Bradley (New York, 1945), 2:132–33. See also pp. 128–30.

6. "The Influence of the Trading Spirit on the Social and Moral Life in America," *The American Review: A Whig Journal of Politics, Literature, Art and Science* (1 Jan. 1845), reprinted in abridged form in *Ideology and Power in the Age of Jackson*, ed. Edwin C. Rozwenc (Garden City, N.Y., 1964), pp. 48–54.

7. Carl Bode, *The Anatomy of American Popular Culture, 1840–1860* (Berkeley and Los Angeles, 1960), chap. 9, surveys this literature.

8. Alfred P. Chandler, Jr., *The Visible Hand: The Managerial Revolution in American Business* (Cambridge, 1977), contains the most authoritative dis-

cussion of marketing changes. Frank Presbrey, *The History and Development of Advertising* (Garden City, N.Y., 1929), remains, after fifty years, a useful source of information for this period. See also James P. Wood, *The Story of Advertising* (New York, 1958). Ralph M. Hower, *History of Macy's of New York, 1858–1919* (Cambridge, Mass., 1943); Herbert Adams Gibbons, *John Wanamaker*, 2 vols. (New York and London, 1926); Robert W. Twyman, *History of Marshall Field & Co., 1852–1906* (Philadelphia, 1954); and John William Ferry, *A History of the Department Store* (New York, 1960), are all helpful on this subject. Above all, there is Hugh Dalziel Duncan, *Culture and Democracy: The Struggle for Form in Society and Architecture in Chicago and the Middle West during the Life and Times of Louis H. Sullivan* (Totowa, N.J., 1965). This is the only book which has treated shopping as a culture; it contains important and stimulating discussions of Chicago stores, consumers, merchandising methods, and artists, and has an especially valuable commentary on Dreiser. I have benefitted greatly from reading the book, which has influenced my own formulation. See particularly chaps. 11 and 12.

9. "My Afternoons Among the Drug Goods," *Harper's Weekly* 1 (31 Oct. 1856): 689–90.

10. Lillie Hamilton French, "Shopping in New York," *The Century* 61 (Mar. 1901): 651.

11. William D. Howells, *A Woman's Reason* (Boston and New York, 1882), p. 185.

12. See W. D. Howells, *The Rise of Silas Lapham*, intro. by Walter J. Meserve (Bloomington and London, 1971), chaps. 1 and 3, for Lapham's business methods, and the initial interviews between him and his architect about the new house. The most quoted episode in the book is the tragicomic dinner-party scene, chap. 14.

13. W. D. Howells, *A Hazard of New Fortunes*, intro. by Everett Carter (Bloomington and London, 1976), p. 49.

14. Theodore Dreiser, *Sister Carrie* (Cleveland and New York, 1900, 1946), pp. 24–25.

15. Ibid., p. 75. Actually, Carrie is shopping alone while in The Fair, but Drouet joins her shopping expedition in the other stores.

16. Ibid., pp. 391–92, 431, 442.

17. Ibid., p. 556.

18. Theodore Dreiser, *The Titan* (New York, 1914, 1925), pp. 31–38, 56–61, 66–73, for Cowperwood's manipulation of the social scene, his use of possessions, and his passion for art. Consciousness of consumption and expenditure powerfully enters Dreiser's 1925 novel, *An American Tragedy*. For an interesting comment upon this subject see Michael Spindler, "Youth, Class and Consumerism in Dreiser's *An American Tragedy*," *Journal of American Studies* 12 (Apr. 1978): 63–80.

19. R. W. Sexton, *American Commercial Buildings of To-Day* (New York, 1928), pp. 153–54. For a discussion of the changing physical character of stores in the 1920s and a more general view of commercial influences on taste, see

Neil Harris, "Museums, Merchandising, and Popular Taste: The Struggle for Influence," in *Material Culture and the Study of American Life,* ed. Ian M. G. Quimby (New York, 1978), pp. 149–74.

20. John Taylor Boyd, Jr., "The Art of Commercial Display," *Architectural Record* 63 (Jan. 1928): 59.

21. Frederick J. Kiesler, *Contemporary Art Appled to the Store and Its Display* (New York, 1930), p. 79. See also Joseph Mayer, *The Revolution in Merchandise* (New York, 1939); A. T. Fischer, *Window and Store Display: A Handbook for Advertisers* (Garden City, N.Y., 1922); Ely Jacques Kahn, "Designing the Bonwit Teller Store," *Architectural Forum* 53 (Nov. 1930): 571; and Shepard Vogelsang, "Architecture and Trade Marks," *Architectural Forum* 50 (June 1929): 897–900.

22. For advertising techniques in the 1920s see Otis Pease, *The Responsibilities of American Advertising* (New Haven, Conn., 1958); Stuart Ewen, *Captains of Consciousness* (New York, 1976); Ruth Schwartz Cowan, "The 'Industrial Revolution' in the Home: Household Technology and Social Change in the 20th Century," *Technology and Culture* 17 (Jan. 1976): 1–23; and an excellent graduate seminar paper, Stephen Freedman, "Corporate Reach and the Family: The Social Context of Food Advertising in National Magazines, 1919–1929" (Chicago, 1978).

23. Floyd W. Parsons, "The New Day in Salesmanship," *Saturday Evening Post* 192 (4 June 1921): 28. Ewen, *Captains of Consciousness* (note 22), describes some of the manipulative strategies. Walter Dill Scott, *The Psychology of Advertising* (Boston, 1921), was an influential text for students of the new methods.

24. Although this subject is not effectively brought together anywhere, there are suggestive hints in Margaret Farrand Thorp, *America at the Movies* (New Haven, Conn., 1939), chaps. 3 and 4; Leo C. Rosten, *Hollywood: The Movie Colony, the Movie Makers* (New York, 1941), chaps. 3 and 9; Hortense Powdermaker, *Hollywood, the Dream Factory: An Anthropologist Looks at the Movie-Makers* (Boston, 1950); the introduction and several of the articles anthologized in *Culture and Commitment, 1929–1945,* ed. Warren Susman (New York, 1972); Robert Sklar, *Movie-Made America: A Social History of American Movies* (New York, 1975), chaps, 8, 13, 14; and Garth Jowett, *Film: The Democratic Art* (Boston, Toronto, 1976), chap. 11.

25. See the homes documented in Arthur Knight and Eliot Elisofson, *The Hollywood Style* (New York, London, 1969), a cultural artifact of its own.

26. Dreyfuss, Bel Geddes, Loewy, and Teague are among the designers who published books on their own. The most useful recent survey on the movement is Donald J. Bush, *The Streamlined Decade* (New York, 1975), concentrating on the 1930s, when the most important work was accomplished. There were, however, some earlier intimations, in architecture and the decorative arts. See also Sheldon and Martha Cheney, *Art and the Machine* (New York, 1936); and Martin Greif, *Depression Modern: The Thirties Style in America* (New York, 1975).

27. Bush, *Streamlined Decade* (note 26) and Greif, *Depression Modern* (note 26)

contain material on the fairs. See also Harris, "Museums, Merchandising, and Popular Taste" (note 19).

28. Alfred P. Sloan, Jr., *My Years with General Motors* (Garden City, N.Y., 1963), chap. 9, describes the introduction of the annual model change. Alfred D. Chandler, Jr., *Strategy and Structure: Chapters in the History of the Industrial Enterprise* (Cambridge, Mass., 1962). See also Daniel J. Boorstin, *The Americans: The Democratic Experience* (New York, 1973), pp. 546–55. This volume also has discussions of the department store (pp. 101–9) and many references to packaging, advertising, and merchandising in the twentieth century.

29. Mary E. Hoffman, *The Buying Habits of Small-Town Women* (Kansas City, Chicago, Rock Island, Atlanta, New York, 1926). The author was director of research for the Ferry-Yanley Advertising Company.

30. George Jean Nathan and H. L. Mencken, *The American Credo: A Contribution Toward the Interpretation of the National Mind* (New York, 1921), pp. 25, 28.

31. Lewis Mumford, "The City," in *Civilization in the United States: An Inquiry by Thirty Americans*, ed. Harold E. Stearns (New York, 1922), p. 9.

32. *The Portable F. Scott Fitzgerald*, intro. by John O'Hara (New York, 1945), from the unpaginated introduction.

33. John O'Hara, *Appointment in Samarra* (New York 1934, 1946), passim.

34. F. Scott Fitzgerald, *The Great Gatsby* (New York, 1925, 1953), pp. 93-94.

35. F. Scott Fitzgerald, "May Day," *Babylon Revisited and Other Stories* (New York, 1960), p. 27.

36. Ibid., p. 32.

37. F. Scott Fitzgerald, *Tender is the Night* (New York, 1933, 1963), p. 55. See also p. 97.

38. Sinclair Lewis, *Main Street* (New York, 1920, 1948), preface page.

39. Ibid., p. 37. The tour is described pp. 33–38. See also chaps. 5 and 17, although consumer concerns touch practically every part of the book.

40. Sinclair Lewis, *Babbitt* (New York, 1922, 1961), pp. 320–21. Schorer wrote the afterword for this Signet edition.

41. Sinclair Lewis, *Babbitt* (New York, 1922, 1950), p. 75.

42. Ibid., p. 95.

43. Ibid., pp. 100–101.

44. Sinclair Lewis, *Dodsworth* (New York, 1929), p. 12.

45. Ibid., p. 58.

46. Ibid., p. 334. For contrast, see the description of Fran, p. 224. For more on buying and specifications of consumer goods, see pp. 183, 187–88, 371, 373.

47. Ibid., pp. 194–95. See also p. 363.

SUGGESTIONS FOR FURTHER READING

Aitken, Hugh G. J. *Taylorism at Watertown Arsenal: Scientific Management in Action, 1908–1915.* Cambridge, Mass., 1960.

———, ed. *The State and Economic Growth.* New York, 1959.

Aldcroft, Derek H., ed. *The Development of British Industry and Foreign Competition, 1875–1914.* Toronto, 1968.

Allen, Zachariah. *The Science of Mechanics as Applied to the Present Improvements in the Useful Arts in Europe and the United States of America.* Providence, 1829.

Althin, Torsten K. W. *C. E. Johansson 1864–1943, The Master of Measurement.* Stockholm, 1948.

Ames, Edward, and Nathan Rosenberg. "The Enfield Armory in Theory and History." *Economic Journal* 78 (1968): 827–42.

Anderson, John. "Report of the Committee on the Machinery of the United States of America." Reprinted in *The American System of Manufactures*, edited by Nathan Rosenberg. Edinburgh, 1969.

Babbage, Charles. *On the Economy of Machinery and Manufactures.* London, 1832. Reprint. New York, 1963.

Bathe, Greville and Dorothy. *Oliver Evans.* Philadelphia, 1936.

Battison, Edwin A. *Muskets to Mass Production.* Windsor, Vt., 1976.

———. "A New Look at the Whitney Milling Machine." *Technology and Culture* 14 (1973): 592–98.

———. "Eli Whitney and the Milling Machine." *Smithsonian Journal of History* 1 (Summer 1966): 9–34.

Benet, Stephen V., ed. *A Collection of Annual Reports and Other Important Papers, Relating to the Ordnance Department.* 3 vols. Washington, D.C., 1878–90.

Berger, Max. *The British Traveller in America, 1836–1860.* New York, 1943.

Birkhimer, William E. *Historical Sketch of the Organization, Administration, Materiel and Tactics of the Artillery, United States Army.* Washington, D.C., 1884. Reprint. New York, 1968.

Bishop, J. L. *History of American Manufacturing*. 3 vols. 3rd ed. New York, 1966.

Blackmore, Howard L. "Colt's London Armory." In *Technological Change: The United States and Britain in the Nineteenth Century*, edited by S.B. Saul. London, 1970.

Blanc, Honoré. *Mémoire important sur la fabrication des armes de guerre, à l'Assemblée Nationale*. Paris, 1790.

Bond, George M. "Standards of Length Applied to Gauge Dimensions." *Journal of the Franklin Institute* 117 (1884): 379.

Bourne, Frederick G. "American Sewing-Machines." In *One Hundred Years of American Commerce*, edited by Chauncey M. Depew, vol. 2. New York, 1895.

Brady, Dorothy. "Relative Prices in the Nineteenth Century." *Journal of Economic History* 24 (1964): 147–48.

————, ed. *Output, Employment and Productivity in the U.S. after 1800*. National Bureau of Economic Research, Studies in Income and Wealth, vol. 30. New York, 1966.

Brandon, Ruth. *A Capitalist Romance: Singer and the Sewing Machine*. Philadelphia, 1977.

Braverman, Harry. *Labor and Monopoly Capital: The Degradation of Work in the Twentieth Century*. New York, 1974.

Broehl, Wayne G. *Precision Valley: The Machine Tool Companies of Springfield, Vermont*. Englewood Cliffs, N.J., 1959.

Broude, Henry W. "The Role of the State in American Economic Development, 1820–1890." In *The State and Economic Growth*, edited by Hugh G. J. Aitken. New York, 1959.

"Brown & Sharpe Manufacturing Company." *Scientific American*, n.s. 41 (1 Nov. 1879): 273.

Brown, Howard Francis. "The Saga of Brown and Sharpe (1833–1968)." Master's thesis, University of Rhode Island, 1971.

Brownlie, D. "John George Bodmer: His Life and Work." *Newcomen Society Transactions* 6 (1925–26): 86–110.

Buchanan, Robertson. *Practical Essays on Mill Work*. 3d ed., rev. London, 1841.

Buckingham, Earle. *Principles of Interchangeable Manufacture*. 2nd ed. New York, 1941.

Burlingame, Luther D. "The Development of Interchangeable Manufacture." *American Machinist* 47 (1914): 295-97.

————. "Pioneer Steps Toward the Attainment of Accuracy." *American Machinist* 47 (1914): 237–43.

Burn, D. L. "The Genesis of American Engineering Competition, 1850–1870." *Economic History* 6 (1931): 292-311.

Buttrick, John. "The Inside Contract System." *Journal of Economic History* 12 (1952): 205–21.

Cain, Louis P., and Paul J. Uselding, eds. *Business Enterprise and Economic Change.* Kent, Ohio, 1973.

Cesari, Gene S. "American Arms-Making Machine Tool Development, 1798–1855." Ph.D. dissertation, University of Pennsylvania, 1970.

Chandler, Alfred D., Jr. *Strategy and Structure: Chapters in the History of the Industrial Enterprise.* Cambridge, Mass., 1962.

————. *The Visible Hand: The Managerial Revolution in American Business.* Cambridge, Mass., 1977.

————. "The Railroads: Pioneers in Modern Corporate Management." *Business History* 39 (1965): 16–40.

————. "The Structure of American Industry in the Twentieth Century: A Historical Overview." *Business History Review* 43 (1966): 255–98.

Chubb, John. "On the Construction of Locks and Keys," and discussion by John Farey. Institution of Civil Engineers, *Proceedings* 9 (1849–50): 331–32.

Clark, Victor S. *History of Manufactures in the United States.* 2 vols. Washington, D.C., 1929.

Clements, P. *Marc Isambard Brunel.* London, 1970.

Coats, A. W., and Ross M. Robertson, eds. *Essays in American Economic History.* London, 1969.

Colvin, Fred H. *Sixty Years with Men and Machines.* New York, 1947.

Conference on Research in Income and Wealth. *Trends in the American Economy in the 19th Century.* Princeton, 1960.

Cooling, B. Franklin, ed. *War, Business, and American Society: Historical Perspectives on the Military-Industrial Complex.* Port Washington, N.Y., 1977.

Cooper, Grace Rogers. *The Sewing Machine: Its Invention and Development.* Washington, D.C., 1976.

Copley, Frank B. *Frederick W. Taylor, Father of Scientific Management.* 2 vols. New York, 1923.

Cumbler, John T. "Labor, Capital and Community: The Struggle for Power." *Labor History* 15 (1974): 395–415.

Dalzell, Robert F., Jr. *American Participation in the Great Exhibition of 1851.* Amherst College Honors Thesis No. 1. Amherst, Mass., 1960.

Daniels, George. "The Big Questions in the History of American Technology." *Technology and Culture* 11 (1970): 1–35.

David, P. A. *Technical Choice, Innovation and Economic Growth: Essays on American and British Experiences in the Nineteenth Century.* Cambridge, 1975.

Davies, Robert B. "International Operations of the Singer Manufacturing Company, 1854–1895." Ph.D. dissertation, University of Wisconsin, 1966.

————. *Peacefully Working to Conquer the World: The Singer Sewing Machine Company in Foreign Markets, 1854–1920.* New York, 1976.

Davis, Lance, et al. *American Economic Growth.* New York, 1972.

Dawley, Alan. *Class and Community: The Industrial Revolution in Lynn.* Cambridge, Mass., 1976.

DeCaindry, William A. *A Compilation of the Laws of the United States Relating to and Affecting the Ordnance Department.* Washington, D.C., 1872.

Depew, Chauncey M., ed. *One Hundred Years of American Commerce.* 2 vols. New York, 1895.

Devinat, Paul. *Scientific Management in Europe.* International Labour Office, Studies and Reports, Series B (Economic Conditions), No. 17. Geneva, 1927.

Dewhurst, P. C. "The Norris Locomotive." Railway and Locomotive Historical Society, *Bulletin* 79 (1950): 1–80.

Deyrup, Felicia J. *Arms Makers of the Connecticut Valley: A Regional Study of the Economic Development of the Small Arms Industry, 1790–1870.* Smith College Studies in History 33. Northampton, Mass., 1948.

Dickenson, H. W. "Joseph Bramah and His Inventions." *Newcomen Society Transactions* 22 (1941–42): 169–86.

———. "Origin and Manufacture of Wood Screws." *Newcomen Society Transactions* 22 (1941–42): 79–89.

———. "Richard Roberts, His Life and Inventions." *Newcomen Society Transactions* 25 (1945–47): 123–37.

Dixie, E. A. "Some More Antique Machine Tools." *American Machinist* 31 (1908): 558–60.

———. "Some Old Gages and Filing Jigs." *American Machinist* 31 (1908): 318.

Donkin, S. B. "Bryan Donkin... 1768–1855." *Newcomen Society Transactions* 27 (1949–50 and 1950–51): 89–95.

Dublin, Thomas. "Women, Work, and Protest in the Early Lowell Mills: 'The Oppressing Hand of Avarice Would Enslave Us.' " *Labor History* 16 (1975): 114.

Duggan, Edward P. "Machines, Markets, and Labor: The Carriage and Wagon Industry in Late-Nineteenth-Century Cincinnati." *Business History Review* 51 (1977): 308–25.

Durfee, William F. "The First Systematic Attempt at Interchangeability in Firearms." *Cassier's Magazine* 5 (1893–94): 469–77.

———. "The History of Interchangeable Construction." American Society of Mechanical Engineers, *Transactions* 14 (1893): 1228.

Earl, Polly Ann. "Craftsmen and Machines: The Nineteenth-Century Furniture Industry." *Technological Innovation and the Decorative Arts,* edited by Ian M. G. Quimby and Polly Ann Earl. Charlottesville, Va., 1974.

Easterlin, Richard A. "The American Population." In Lance Davis et al., *American Economic Growth.* New York, 1972.

Eddy, C. W. "American Implements and Methods of Economizing Labor." *The Journal of the Royal Agricultural Society of England* 20 (1859): 109–131.

"Edward Clark." *Obituary Record of Donors and Alumni of Williams College, 1882–3,* No. 18 (1883): 304–6.

Emerson, Harrington. *Efficiency as a Basis for Operations and Wages.* New York, 1911.

Ewen, Stuart. *Captains of Consciousness: Advertising and the Social Roots of the Consumer Culture.* New York, 1976.

Ezell, Edward C. "The Development of Artillery for the United States Land Service before 1861: With Emphasis on the Rodman Gun." Master's thesis, University of Delaware, 1963.

————. "James Henry Burton et le transfert du système américain aux arsenaux du gouvernement impérial russe." *Le Musée d'Armes* (Liège) *Bulletin* 5 (1977): 13–20; paper delivered in English at the 20th Anniversary Conference of the Hagley Program, Eleutherian Mills-Hagley Foundation, Wilmington, Delaware.

Faler, Paul. "Cultural Aspects of the Industrial Revolution: Lynn, Massachusetts, Shoemakers and Industrial Morality." *Labor History* 15 (1974): 367–94.

Falk, Stanley L. "Soldier-Technologist: Major Alfred Mordecai and the Beginnings of Science in the United States Army." Ph.D. dissertation, Georgetown University, 1959.

Ferguson, Eugene S. *Early Engineering Reminiscences (1815–1840) of George Escol Sellers.* Washington, D.C., 1965.

————. "On the Origin and Development of American Mechanical 'Know-How.'" *Midcontinent American Studies Journal* 3 (1962): 3–15.

Fernie, John. "Decimal System of Measurement with Description of Mr. Whitworth's Measuring Machine and Standard Gauges." *Proceedings of the Institution of Mechanical Engineers* (1859): 110–22.

Ferry, John William. *A History of the Department Store.* New York, 1960.

Fisher, Marvin W. *Workshops in the Wilderness: The European Response to American Industrialization, 1830–1860.* New York, 1967.

Fishlow, Albert. "Comparative Consumption Patterns, the Extent of the Market, and Alternative Development Strategies." In *Micro-Aspects of Development,* edited by Eliezer Ayal. New York, 1973.

Fitch, Charles H. "Report on the Manufactures of Interchangeable Mechanism." In U.S. Census Office, *Report on the Manufactures of the United States at the Tenth Census, 1880,* vol. 2, pp. 611–704. Washington, D.C., 1883.

————. "The Rise of A Mechanical Ideal." *Magazine of American History* 11 (1884): 516–27.

Floud, R.C. "The Adolesence of American Engineering Competition, 1860–1900." *Economic History Review,* 2d ser. 27 (1974): 57–71.

Ford, Henry. "Mass Production." In *Encyclopaedia Britannica,* 13th ed., suppl. vol. 2, pp. 821–23. London, 1926.

Fries, Russell I. "British Response to the American System: The Case of the

Small-Arms Industry after 1850." *Technology and Culture* 16 (1975): 377–403.

Gallman, Robert. "Commodity Output, 1839–1899." In Conference on Research in Income and Wealth, *Trends in the American Economy in the 19th Century*. Princeton, 1960.

———. "The Record of American Economic Growth." In Lance Davis et al., *American Economic Growth*. New York, 1972.

———, ed. *Recent Development in the Study of Business and Economic History: Essays in Memory of Herman E. Kroos*. Greenwich, Conn., 1977.

Gibb, George S. *The Saco-Lowell Shops: Textile Machinery Building in New England, 1813–1949*. Cambridge, Mass., 1950.

Gilbert, K. R. *The Portsmouth Blockmaking Machinery*. London, 1956.

———. "Henry Maudslay 1771–1831." *Newcomen Society Transactions* 44 (1971–72): 49–62.

Gitelman, Howard. "The Labor Force at Waltham Watch during the Civil War." *Journal of Economic History* 25 (1965): 214–44.

———. "Occupational Mobility within the Firm." *Industrial and Labor Relations Review* 20 (1966): 50–63.

———. "Perspectives on Industrial Violence." *Business History Review* 47 (1973): 1–23.

———. *Workingmen of Waltham*. Baltimore, 1974.

Goodrich, Carter. "Internal Improvements Reconsidered." *Journal of Economic History* 30 (1970): 289–311.

Great Britain. Board of Ordnance. *Report of the Committee on the Machinery of the United States of America*. Parliamentary Papers, 1854–55, vol. 50. London, 1855.

———. Parliament. House of Commons. *Report of the Select Committee on Small Arms*. Parliamentary Papers, 1854, vol. 18. London, 1854.

———. *New York Industrial Exhibition: Special Reports of Mr. George Wallis and Mr. Joseph Whitworth*. Parliamentary Papers, 1854, vol. 26. London, 1854.

———. *Select Committee on Exportation of Machinery*. Parliamentary Papers, 1841, vol. 7, Second Report. London, 1841.

Habakkuk, H. J. *American and British Technology in the Nineteenth Century: The Search for Labour-Saving Inventions*. Cambridge, 1962.

Halsey, Frederick A. *Methods of Machine Shop Work*. New York, 1914.

———. "The National Cash Register Company's Experiment." *American Machinist* 24 (1901): 688–89.

Harris, Neil. *The Artist in American Society: The Formative Years, 1790–1860*. New York, 1966.

———. "Museums, Merchandising, and Popular Taste: The Struggle for Influence." In *Material Culture and the Study of American Life*, edited by Ian M. G. Quimby. New York, 1978.

Hatch, Alden. *Remington Arms in American History.* New York, 1956.

Haven, Charles T., and Frank A. Belden. *A History of the Colt Revolver.* New York, 1940.

Hazard, Blanche Evans. *The Organization of the Boot and Shoe Industry in Massachusetts before 1875.* Cambridge, Mass., 1921.

Hill, Forest G. "Formative Relations of American Enterprise, Government and Science." *Political Science Quarterly* 75 (1960): 400–19.

————. "Government Engineering Aid to Railroads before the Civil War." *Journal of Economic History* 11 (1951): 235–46.

————. *Roads, Rails and Waterways: The Army Engineers and Early Transportation.* Norman, Okla., 1957.

Holt, James. "Trade Unionism in the British and American Steel Industries, 1880–1914: A Comparative Study." *Labor History* 18 (1977): 5–35.

Horwitz, Richard P. *Anthropology Toward History: Culture and Work in a 19th-Century Maine Town.* Middletown, Conn., 1978.

Hounshell, David A. "From the American System to Mass Production: The Development of Manufacturing Technology in the United States, 1850–1920." Ph.D. dissertation, University of Delaware, 1978.

Howard, Robert A. "Interchangeable Parts Reexamined: The Private Sector of the American Arms Industry on the Eve of the Civil War." *Technology and Culture* 19 (1978): 633–49.

Hubbard, Guy. "Development of Machine Tools in New England." *American Machinist* 59 (1923), 60 (1924), 61 (1924).

Hughes, Thomas P. "Inventors: The Problems They Choose, the Ideas They Have, and the Inventions They Make." In *Technological Innovation: A Critical Review of Current Knowledge,* edited by Patrick Kelly and Melvin Kranzberg. San Francisco, 1978.

"(The) Influence of the Trading Spirit on the Social and Moral Life in America." *The American Review: A Whig Journal of Politics, Literature, Art, and Science* 1 (1845); reprinted in abridged form in *Ideology and Power in the Age of Jackson,* edited by Edwin C. Rozwenc. Garden City, N.Y., 1964.

Jenkins, Reese V. *Images and Enterprise: Technology and the American Photographic Industry 1839 to 1925.* Baltimore, 1975.

Jeremy, David J. "British Textile Technology Transmission to the United States." *Business History Review* 47 (1973): 24–52.

————. "Innovation in American Textile Technology during the Early 19th Century." *Technology and Culture* 14 (1973): 40–76.

————. "The Transmission of Cotton and Woollen Manufacturing Technologies between Britain and the U.S.A. from 1790 to the 1830s." Ph.D. dissertation, London School of Economics and Political Science, University of London, 1978.

Kammen, Michael. "From Liberty to Prosperity: Reflections upon the Role of Revolutionary Iconography in National Tradition." *Proceedings of the Ameri-*

can *Antiquarian Society* 86 (1976): 263–72.

Koht, Haldvan. *The American Spirit in Europe: A Survey of Transatlantic Influences*. Philadelphia, 1949.

Kouwenhoven, John A. *Made in America: The Arts in Modern Civilization*. Newton Centre, Mass., 1948.

————. "Waste Not, Have Not: A Clue to American Prosperity." *Harper's Magazine* 218 (Mar. 1959): 72–79, 81.

Landes, David. *The Unbound Prometheus: Technological Change and Industrial Development in Western Europe from 1750 to the Present Day*. Cambridge, 1969.

Leaver, E.W., and J.J. Brown, "Machines without Men." *Fortune* 34 (Nov. 1946): 165.

Lebergott, Stanley. "Labor Force and Employment, 1800–1960." In *Output, Employment and Productivity in the U.S. after 1800*, ed. Dorothy Brady. National Bureau of Economic Research, Studies in Income and Wealth, vol. 30. New York, 1966.

————. *Manpower in Economic Growth*. New York, 1962.

Leiby, James. *Carroll Wright and Labor Reform*. Cambridge, Mass., 1960.

Leland, Ottilie M. *Master of Precision: Henry M. Leland*. Detroit, 1966.

Liggett, John V. *Fundamentals of Position Tolerance*. Dearborn, Mich., 1970.

Littauer, S. B. "Development of Statistical Quality Control in the United States." *American Statistician* 4 (1950): 14–20.

Lively, Robert A. "The American System: A Review Article." *Business History Review* 29 (1955): 81–96.

Long, Clarence. *Wages and Earnings in the United States 1860–1890*. Princeton, 1960.

Lyon, Peter. "Isaac Singer and His Wonderful Sewing Machine." *American Heritage* 9 (Oct. 1958): 34–39, 103–9.

McKenzie, Fred A. *The American Invaders*. London, 1901.

McLaurin, Melton Alonza. *Paternalism and Protest: Southern Cotton Mill Workers and Organized Labor, 1875–1905*. Westport, Conn., 1971.

McNeil, Ian. *Joseph Bramah*. Newton Abbot, 1968.

Maier, Charles S. "Between Taylorism and Democracy: European Ideologies and the Vision of Industrial Productivity in the 1920s." *Journal of Contemporary History* 5 (1970): 27–61.

Martin, Edgar W. *The Standard of Living in 1860: American Consumption Levels on the Eve of the Civil War*. Chicago, 1942.

Maudslay, C. C. *Henry Maudslay*. London, 1948.

Meier, Hugo. "Technology and Democracy, 1800–1860." *Mississippi Valley Historical Review* 43 (1957): 618–40.

Mesick, Jane Louise. *The English Traveller in America, 1785–1835.* New York, 1922.

(The) Micrometer's Story, 1867–1902. Providence, R.I., n.d.

Mirsky, Jeanette, and Allan Nevins. *The World of Eli Whitney.* New York, 1952.

Molloy, Peter M. "Technical Education and the Young Republic: West Point as America's École Polytechnique, 1802–1833." Ph.D. dissertation, Brown University, 1975.

Moore, Wayne R. *Foundations of Mechanical Accuracy.* Bridgeport, Conn., 1970.

Mordecai, Alfred. *Report of Experiments on Gunpowder, Made at Washington Arsenal, in 1843 and 1844.* Washington, D.C., 1845.

———. *Second Report of Experiments on Gunpowder, Made at Washington Arsenal, in 1845, '47, and '48.* Washington, D.C., 1849.

Murphy, John J. "Entrepreneurship in the Establishment of the American Clock Industry." *Journal of Economic History* 26 (1966): 169–86.

———. "The Establishment of the American Clock Industry: A Study in Entrepreneurial History." Ph.D. dissertation, Yale University, 1961.

Musson, A. E. "James Nasmyth and the Early Growth of Mechanical Engineering." *Economic History Review*, 2nd ser. 10 (1957–58): 121–27. Reprinted in A. E. Musson and E. Robinson, *Science and Technology in the Industrial Revolution.* Manchester, England, 1969.

———. "Joseph Whitworth and the Growth of Mass-Production Engineering." *Business History* 17 (1975): 109–49.

———. "The 'Manchester School' and Exportation of Machinery." *Business History* 14 (1972): 17–50.

——— and E. Robinson. *Science and Technology in the Industrial Revolution.* Manchester, England, 1969.

Nasmyth, James. *James Nasmyth, Engineer. An Autobiography.* Edited by Samuel Smiles. London, 1883.

———. "Remarks on the Introduction of the Slide Principle in Tools and Machines Employed in the Production of Machinery." In *Practical Essays on Mill Work*, edited by Robertson Buchanan. 3d ed., rev. London, 1841.

Navin, T. R. *The Whitin Machine Works since 1831.* Cambridge, Mass., 1950.

Nelson, Daniel. *Managers and Workers: Origins of the New Factory System in the United States, 1880–1920.* Madison, Wis., 1975.

———. "The New Factory System and the Unions: The NCR Dispute of 1901." *Labor History* 15 (1974): 163–78.

———. "Scientific Management, Systematic Management, and Labor, 1880–1915." *Business History Review* 48 (1974): 479–500.

Nevins, Allan, ed. *America through British Eyes.* New York, 1948.

Nutter, Waldo E. *Manhattan Firearms*. Harrisburg, Pa., 1958.

Ozanne, Robert. *A Century of Labor-Management Relations*. Madison, Wis., 1967.

———. *Wages in Theory and Practice*. Madison, Wis., 1968.

Parkhurst, E. G. "One of the Earliest Milling Machines." *American Machinist* 23 (1900): 217–25.

Parsons, Floyd W. "The New Day in Salesmanship." *Saturday Evening Post* 192 (4 June 1921): 28

Parsons, William B. "Engineering Development of the Far East." *Engineering Magazine* 19 (1900): 481–92.

Pease, Otis. *The Responsibilities of American Advertising: Private Control and Public Influence*. New Haven, Conn., 1958.

Petree, J. F. *Henry Maudslay (1771–1831) and Maudslay, Sons & Field Ltd.* London, 1949.

———. "Henry Maudslay—Pioneer of Precision." In Institution of Mechanical Engineers, *Engineering Heritage*, vol. I. London, 1964.

Porter, Glenn. "Management." In *A History of Technology: The Twentieth Century*, edited by T. I. Williams. New York, 1978.

———, and Harold C. Livesay. *Merchants and Manufacturers: Studies in the Changing Structure of Nineteenth-Century Marketing*. Baltimore, 1971.

Potter, David M. *People of Plenty: Economic Abundance and the American Character*. Chicago, 1954.

Presbrey, Frank. *The History and Development of Advertising*. Garden City, N.Y., 1929.

Pye, David. *The Nature and Art of Workmanship*. Cambridge, 1968.

Reed, C. Wingate, "Decius Wadsworth, First Chief of Ordnance, U.S. Army, 1812–1821." *Army Ordnance* 24 (1943): 527–30; 25 (1943): 113–16.

Reznick, Samuel. "The Rise and Early Development of Industrial Consciousness in the United States, 1760–1830." *Journal of Economic and Business History* 42 (1932): 784–811.

Richards, John. *On the Arrangement, Care, and Operation of Wood-Working Factories and Machinery*. London, 1885.

Robinson, Jesse S. *The Amalgamated Association of Iron, Steel and Tin Workers*. Baltimore, 1920.

Rodman, Thomas Jefferson. *Reports of Experiments on the Properties of Metals for Cannon, and the Qualities of Cannon Powder*. Boston, 1861.

Roe, Joseph W. *English and American Tool Builders*. New Haven, Conn., 1916.

———. "Interchangeable Manufacture." *Newcomen Society Transactions* 17 (1936–37): 165–74.

———. "Machine Tools in America." *Journal of the Franklin Institute* 225 (1938): 499–511.

Roland, Henry. "Six Examples of Successful Shop Management, V." *Engineering Magazine* 12 (1897): 997–98.

Rolt, L. T. C. *Tools for the Job: A Short History of Machine Tools.* London, 1965.

Rosenberg, Nathan. *The American System of Manufactures.* Edinburgh, 1969.

———. "American Technology: Imported or Indigenous?" *American Economic Review Papers and Proceedings*, Feb. 1977.

———. "Anglo-American Wage Differentials in the 1820s." *Journal of Economic History* 27 (1967): 221–29.

———. *Perspectives on Technology.* Cambridge, 1976.

———. "Technological Change in the Machine Tool Industry, 1840–1910." *Journal of Economic History* 23 (1963): 414–46; reprinted in idem, *Perspectives on Technology,* chap. 1.

———. *Technology and American Economic Growth.* New York, 1972.

Rosenbloom, Richard. "Men and Machines: Some 19th-Century Analyses of Mechanization." *Technology and Culture* 4 (1964): 489–511.

Rothbarth, E. "Causes of Superior Efficiency of U.S.A. Industry as Compared to British Industry." *Economic Journal* 54 (1946): 383–90.

Saul, S. B. "The Engineering Industry." In *The Development of British Industry and Foreign Competition, 1875–1914,* edited by Derek H. Aldcroft. Toronto, 1968.

———. "The Machine Tool Industry in Britain to 1914." *Business History* 10 (1968): 22–43.

———. "The Motor Industry in Britain to 1914." *Business History* 5 (1962): 22–44.

———, ed. *Technological Change: The United States and Britain in the Nineteenth Century.* London, 1970.

Sawyer, John E. "The Social Basis of the American System of Manufacturing." *Journal of Economic History* 14 (1954): 361–79.

Scheiber, Harry N. "Government and the Economy: Studies of the "Commonwealth' Policy in Nineteenth-Century America." *Journal of Interdisciplinary History* 3 (1972): 135–54.

Scott, John. *Genius Rewarded; or the Story of the Sewing Machine.* New York, 1880.

Scott, Walter Dill. *The Psychology of Advertising.* Boston, 1921.

Sharpe, Henry Dexter. *Joseph R. Brown, Mechanic, and the Beginnings of Brown and Sharpe.* New York, 1949.

Sharpe, Lucian. "Letter of December 10, 1887." Printed as "Development of the

Micrometer Caliper," *American Machinist*, 15 Dec. 1892, pp. 9–10.

Shaw, Ralph R. *Engineering Books Available in America Prior to 1830.* New York, 1933.

Singer, Isaac. "Looking Back: A Bird's Eye View of Our Trade." *Sewing Machine Times*, n.s. 18 (25 Oct. 1908): 8.

Singer Manufacturing Company. *Catalogue of Singer Sewing-Machine.* New York, 1896.

———. *Mechanics of the Sewing Machine.* New York, 1914.

Sloan, Alfred P., Jr. *My Years with General Motors.* Garden City, N.Y., 1963.

Smith, L. R. "We Build a Plant to Run without Men." *Magazine of Business* 55 (Feb. 1929): 135–38, 200.

Smith, Merritt Roe. "Eli Whitney and the American System of Manufacturing." In *Technology in America: A History of Individuals and Ideas,* edited by Carroll W. Pursell. Cambridge, Mass., 1981.

———. *Harpers Ferry Armory and the New Technology: The Challenge of Change.* Ithaca, N.Y., 1977.

———. "John Hall, Simeon North, and the Milling Machine: The Nature of Innovation among Antebellum Arms Makers." *Technology and Culture* 14 (1973): 573–91.

———. "Military Arsenals and Industry Before World War I." In *War, Business, and American Society: Historical Perspectives on the Military Industrial Complex,* edited by B. Franklin Cooling. Port Washington, N.Y., 1977.

Spalding, B. F. "The 'American System' of Manufacture." *American Machinist* 13 (1890): 11.

Stearns, Charles. *The National Armories.* Springfield, 1852.

Steeds, W. *A History of Machine Tools, 1700–1910.* Oxford, 1969.

Strassman, W. Paul. *Risk and Technological Innovation.* Ithaca, N.Y., 1958.

Sweet, John E. *Machine Practice.* Ithaca, N.Y., 1875.

Taylor, Frederick Winslow. *Scientific Management, Comprising Shop Management, The Principles of Scientific Management, Testimony before the Special House Committee.* New York, 1947.

Temin, Peter. *Causal Factors in American Economic Growth in the Nineteenth Century.* London, 1975.

———. *Iron and Steel in the Nineteenth Century: An Economic Inquiry.* Cambridge, Mass., 1964.

———. "Labor Scarcity and the Problem of American Industrial Efficiency in the 1850's." *Journal of Economic History* 26 (1966): 277–98.

United States. Department of Commerce and Labor, Bureau of Manufactures. *Monthly Consular and Trade Reports,* No. 317 (Feb. 1907, Doc. ser. no. 5176), pp. 60–61.

————. Ordnance Department. *Regulations of the Government of the Ordnance Department.* Washington, D. C., 1834.

————. Ordnance Department. *Reports of Experiments on the Strength and Other Properties of Metals for Cannon.* Philadelphia, 1856.

Ure, Andrew. *Philosophy of Manufactures.* London, 1835.

Urwick, L., and E. F. L. Brech. *The Making of Scientific Management.* Vol. 1: *Thirteen Pioneers.* London, 1951.

Uselding, Paul. "An Early Chapter in the Evolution of American Industrial Management." In *Business Enterprise and Economic Change,* edited by Louis P. Cain and Paul J. Uselding. Kent, Ohio, 1973.

————. "Elisha K. Root, Forging, and the 'American System.'" *Technology and Culture* 15 (1974): 543–68.

————. "Henry Burden and the Question of Anglo-American Technological Transfer in the Nineteenth Century." *Journal of Economic History* 30 (1970): 312–37.

————. "Studies of Technology in Economic History." In *Recent Development in the Study of Business and Economic History: Essays in Memory of Herman E. Kroos,* edited by Robert Gallman. Greenwich, Conn., 1977.

Van Slyck, J. D. *New England Manufacturers and Manufacturing.* 2 vols. Boston, 1879.

Veblen, Thorstein. *The Instinct of Workmanship.* 1914. Reprint. New York, 1964.

Wade, William. "Early Systems of Artillery." *Ordnance Notes, No. 25,* pp. 140–41. Washington, D. C., 1874.

Wagoner, Harless D. *The U.S. Machine Tool Industry from 1900–1950.* Cambridge, Mass., 1968.

Walkowitz, Daniel J. "Statistics and the Writing of Working Class Culture: A Statistical Portrait of the Iron Workers in Troy, New York, 1860–1880." *Labor History* 15 (1974): 427.

Wallace, Anthony F. C. *Rockdale: The Growth of an American Village in the Early Industrial Revolution.* New York, 1978.

Ware, Norman. *The Industrial Worker, 1840–60.* Boston, 1924.

Webster, Ambrose. "Early American Steel Rules." *American Machinist* 17 (1894): 7.

Wheeler & Wilson Sewing Machine Company. *The Sewing Machine: Its Origin, Introduction into General Use, Progress and Extent of Its Manufacture [and] A Great Machine-Shop Described.* New York, 1863.

Whitworth, Joseph. "Special Report of Mr. Joseph Whitworth." Reprinted in *The American System of Manufactures,* edited by Nathan Rosenberg. Edinburgh, 1969.

Wilkins, Mira. *The Emergence of Multinational Enterprise.* Cambridge, Mass.,

1970.

Wilkinson, Norman B. "Brandywine Borrowings from European Technology." *Technology and Culture* 4 (1963): 1–13.

———. "The Forgotten 'Founder' of West Point." *Military Affairs* 24 (1960–61): 177–88.

Williamson, Harold F. *Winchester.* Washington, D.C., 1952.

Wood, James P. *The Story of Advertising.* New York, 1958.

Woodbury, Robert S. *History of the Gear Cutting Machine.* Cambridge, Mass., 1958.

———. *History of the Grinding Machine.* Cambridge, Mass., 1959.

———. *History of the Lathe.* Cambridge, Mass., 1961.

———. *History of the Milling Machine.* Cambridge, Mass., 1960.

———. "The Legend of Eli Whitney and Interchangeable Parts." *Technology and Culture* 1 (1960): 235–53; reprinted in *Essays in American Economic History*, edited by A. W. Coats and Ross M. Robertson (London, 1969) and in *Technology and Culture*, edited by Melvin Kranzberg and William H. Davenport (New York, 1972).

———. *Studies in the History of Machine Tools.* Cambridge, Mass., 1972.

Wright, Carroll D. "Report on the Factory System of the United States." In U.S. Department of the Interior, Census Office, *Tenth Census, 1880, Manufactures.* Washington, D.C., 1883.

INDEX

Page numbers in *italics* refer to illustrations.

Sharpe, *118*

Polhem, Christopher, 25, 29, 44 (n.13), 121

Polhem, Gabriel, 121

"Polhem sticks," 121

Portsmouth Naval Dockyard. *See* Blockmaking machinery

Position tolerance, 104, 123 (n.5), 124 (n.6)

Pratt & Whitney, measuring machine, *120*, 121

Precision manufacture, 103–26; definition, 104

Prescott, Benjamin, 70

Proctor, William F., 135

Providence Tool Co., 88

Quality control, 73, 75. *See also* Inspection procedures

Railroads, expansion of network, 153–54; cooperative shipping arrangements, 153; and line-and-staff concept, 158, 160; and streamlined design, 202

Railroad locomotives, 11, 21 (n.50), 177

Rand Drill Co. *See* Halsey, Frederick A.

Refining, large-batch techniques, 154, 155, 162

Reiss, Hans, 144

Remington & Sons Co., 88, 144, 151 (n.51), 160

Rennie, John, 33

Research laboratories, industrial, 164

Robb, John, 90

Robbins & Lawrence (Windsor, Vt.), xii, 29, 88, 131, 144

Roberts, Richard, 25, 33, 36, 37, 38, 39, 46 (n.45)

Rochefontaine, Stephen, 68

Rodman, Thomas J., 85, 86, 96 (n.9)

Rogers, William A., 120

Rogers-Bond Comparator, *120*, 121

Rohde, Gilbert, 202

Root, Elisha K., 4, 112 (n.24)

Routing slips, 158

Scales and rules, for length mea-surement, 112, 114, 115, 118

Scientific management, 9, 141, 151 (n.51), 158

Screw caliper, 117

"Second Industrial Revolution," 169

"Self-acting" machine tools, 25, 31, 34, 36, *37*, 39, 40, 92

Sellers, George Escol, 34, 46 (n.43)

Sewing-machine industry, 129–52, 156, 166, 177

Sharp, Roberts & Co. (Manchester), 37

Sharps Rifle Co. (Hartford), 88, 131, 132

Shop-order system, 158

Simonson, Lee, 202

Singer & Co. (New York), xvii, 11, 130, 132, 133, 134, 135, 136, 137, 140, 141, 142, 143, 144, 149 (n.31), 163

Single-purpose machines. *See* Special-purpose machines

Size accuracy, 104

Slater, Samuel, xi

Smeaton, John, 32

Smiles, Samuel, 37

A.O. Smith Corp. (Milwaukee), 9

Smith, Adam, 8, 52

Smith, William, 144, 145

Soap, mass production of, 154, 162

Solid caliper ("snap") gauges, 112

Special-purpose machines, xii, 4, 5, 13 (n.13), 42, 52, 58

Springfield Armory, 16 (n.10), 29, 63–102, *81*, 112, 115, 124 (n.6), 125 (n.24), 128, 130, 136, 137, 156

Standardization, 34, 35, 36, 37, 39, 40, 47 (n.56, 57), 52, 58, 69, 74, 83, 86, 119, 153, 154, 156

Standards, 40, 104, 106, 112, 114, 123 (n.5), 144, 145, 146

Starr, Nathan, 76

Steam engines, 161, 172

Steelmaking, 155

Stock-trail system, 67, 69, 97 (n.12)

Stowe, Harriet Beecher, 8

Stubblefield, James, 70, 71, 72, 79

Sugar processing, 154

Sweet, John E., 119

Sweet's measuring machine, 119, *120*

Swift, Gustavus F., 162–63
Syracuse Twist Drill Co., 119
Système Palmer. *See* Screw caliper

Talcott, George, 80, 81, 82, 92, 93
Taylor, Frederick Winslow, 10, 20
  (n.43, 44), 151 (n.51), 155, *157*, 158,
  159–60
Taylorism, 9, 12, 141
Taylor, Walter, and Southampton
  blockmaking machinery, 34
Teague, Walter Dorwin, 202
Technological convergence, 89, 107
Telegraph network, expansion, 153–54
Telephone equipment, 163, 166
Terry, Eli, xi, 20 (n.45)
Textile manufacture, 4–5, 173
Thomson, J. Edgar, 170 (n.4)
Time-and-motion studies, 158
Time clocks, 93
Tobacco products, 162
Tocqueville, Alexis de, 191
Tolerances, 103–26; definition, 104
Topographical Bureau, U.S. Army,
  63, 64
Tousard, Louis de, 66, 97 (n.9), 98
  (n.10, 12)
Towne, Henry A., *157*
Trade unions, 175
Treschler, Christopher, of Dresden,
  117
"True plane," 111
Tyler, Daniel, 79, 80

Uniformity system, of U.S. Ordnance
  Department, as predecessor to interchangeability, 66, 67, 68, 69, 70,
  71, 72, 73, 74, 79, 80, 82, 83, 87,
  91, 94, 97 (n.12)
United Fruit Co., 163
Ure, Andrew, 8, 34, 36

Veblen, Thorstein, 19 (n.29)
Vernier calipers, 115
Vincennes Arsenal, 67
Volume production, 153

Wade, William, 84, 85
Wadsworth, Decius, 68, 69, 70, 71,
72, 74, 82, 98 (n.16)
Walbach, Louis A.B., 85, 101 (n.48)
Waltham Watch Co., 178, 179
Watch manufacture, 156, 161, 172
Watervliet Arsenal, 80
Watt, James, 32; shop micrometer,
  114
Webster, Ambrose, 112, 115
Wedgwood, Josiah, 32
Wheeler & Wilson Manufacturing Co.
  (Nathaniel Wheeler and Allen B.
  Wilson), Bridgeport, 130, 131, 132,
  133, 135, 136, 137, 144, 148 (n.11)
Whitney, Eli, xi, 2, 4, 15 (n.3), 25,
  27, 44 (n.9), 68, 70, 98 (n.15, 16),
  144
Whittemore, Amos, 21 (n.49)
Whitworth, Joseph, 7, 21 (n.51), 23
  (n.63), 25, 31, 33, 36, 37, 39, 41, 42,
  43, 44 (n.7), 46 (n.74), 48 (n.66),
  111, 128, 147 (n.3); measuring ma-
  chine, 111; plug-and-ring gauges, 46
  (n.47), 112, *113;* standards, 40
Wickham, Marine T., 76
Wilkinson, John, 32, 33
Willcox & Gibbs Sewing Machine Co.
  (James A. Willcox and James A
  Gibbs), 130, 131, 132, 137, 148
  (n.17)
Wilson, James, 130
Winans, Thomas, 21 (n.50)
Winchester Repeating Arms Co., 144,
  151 (n.51)
Wolcott, Oliver, 68
Wood, Jethro, 55
Woodworking machinery, *56*, 83, 100
  (n.44). *See also* American system
  of manufactures, in woodworking
  industry
Workers: armory, 90–94; in factories
  employing American system,
  171–88
Wright, Carroll D., 171, 172, 179,
  181, 183, 184
Wyatt brothers, wood-screw factory
  (Staffordshire), 34
Wyke, John, 32, 33

Yale & Towne. *See* Towne, Henry A.